Studies in Big Data

Volume 92

Series Editor

Janusz Kacprzyk, Polish Academy of Sciences, Warsaw, Poland

The series "Studies in Big Data" (SBD) publishes new developments and advances in the various areas of Big Data- quickly and with a high quality. The intent is to cover the theory, research, development, and applications of Big Data, as embedded in the fields of engineering, computer science, physics, economics and life sciences. The books of the series refer to the analysis and understanding of large, complex, and/or distributed data sets generated from recent digital sources coming from sensors or other physical instruments as well as simulations, crowd sourcing, social networks or other internet transactions, such as emails or video click streams and other. The series contains monographs, lecture notes and edited volumes in Big Data spanning the areas of computational intelligence including neural networks, evolutionary computation, soft computing, fuzzy systems, as well as artificial intelligence, data mining, modern statistics and Operations research, as well as self-organizing systems. Of particular value to both the contributors and the readership are the short publication timeframe and the world-wide distribution, which enable both wide and rapid dissemination of research output.

The books of this series are reviewed in a single blind peer review process.

Indexed by SCOPUS, SCIMAGO and zbMATH.

All books published in the series are submitted for consideration in Web of Science.

More information about this series at http://www.springer.com/series/11970

Umang Singh · Ajith Abraham ·
Arturas Kaklauskas · Tzung-Pei Hong
Editors

Smart Sensor Networks

Analytics, Sharing and Control

 Springer

Editors
Umang Singh
Institute of Technology and Science
Ghaziabad, Uttar Pradesh, India

Arturas Kaklauskas
Department of Construction
Management and Real Estate
Vilnius Gediminas Technical University
Vilnius, Lithuania

Ajith Abraham
Scientific Network for Innovation
and Research Excellence
Machine Intelligence Research Labs
(MIR Labs)
Auburn, WA, USA

Tzung-Pei Hong 🆔
Department of Computer Science
and Information Engineering
AI Research Center
National University of Kaohsiung
Kaohsiung, Taiwan

ISSN 2197-6503   ISSN 2197-6511   (electronic)
Studies in Big Data
ISBN 978-3-030-77216-1   ISBN 978-3-030-77214-7   (eBook)
https://doi.org/10.1007/978-3-030-77214-7

This Springer imprint is published by the registered company Springer Nature Switzerland AG
The registered company address is: Gewerbestrasse 11, 6330 Cham, Switzerland

Preface

Rapid use of digital services and continual increase of Internet users are demanding a pervasive and ubiquitous connectivity environment across the globe. Due to this, smart wireless sensors are getting tremendous popularity in real-life applications like smart agriculture, smart health care, smart city, etc. Smart networks present the next evolutionary development where sensory data is collected from a different variety of networks of interconnected sensors from heterogeneous distributed locations. Smart sensors are tiny devices that gather information and also measure it from different channels and adjust the data to support processing and decision making to enhance network performance. Smart sensor networks are being considered the most significant technologies as a futuristic vision for custom-specific architectures, user design, monitoring and controlling of information with enhanced security and minimum power consumption. In today's scenario, different latest devices and appliances are connecting through Internet and moving one step ahead in the era of IoT to form intelligent smart sensor networks for finding optimal ways to avail information in complex environments. These networks formed intelligent connection among devices and present a new kind of world to fulfill demands such as high degree of intelligence, interconnection and network technologies. The demand of smart sensors is increasing day by day for decision making, monitoring and controlling of sensitive information as compared to traditional sensors. Further, to meet the demands of users, a smart sensor network requires proper management of data and enlightens attention to serious concerns of data collection efficiency, data sharing to intended users, data analytics, privacy and security.

To form an efficient network, there is a need to ensure and incorporate AI-based techniques and machine learning-based algorithms to develop intelligent computation and monitoring of system environments to improve functionality and utility aspects of a system.

This edited volume entitled *Smart Sensor Networks: Analytics, Sharing and Control* is a timely contribution to the smart sensor networks that are gaining commercial interest and momentum in the development of social and intelligent applications. This provides detailed coverage on smart sensors and devices, types of sensors, data analysis and monitoring with the help of smart sensors, data processing, decision making, impact of machine learning algorithms and artificial intelligence-related methodologies for data analysis and understanding of machine learning-based social and intelligent applications in smart sensor networks. The objectives of this book are as follows:

- To present a valuable insight into the original research and review articles on the latest achievements that will contribute to the field of smart sensor networks and their usage in related domain.
- To analyze and present the state of the art of the smart sensors AI-related technologies and methodologies.
- To understand depth insight of network techniques, design and principles of smart sensor networks.
- To highlight on survey of technologies employed to smart sensor networks, data collection, data monitoring, control and management.
- To discuss research-allied challenges of interesting real-life applications like smart city, smart agriculture, e-health care and smart social sensing networks, etc.

The book comprises nine chapters and is arranged as follows: In Chapter "An Overview of Artificial Intelligence Technology Directed at Smart Sensors and Devices from a Modern Perspective," Monteiro et al. provided an overview of artificial intelligence technology directed at smart sensors and devices. Chapter "The Role of Smart Sensors in Smart City" presents an overview of smart sensors and highlights the need of smart sensors in smart cities for remote control technologies. Harpreet et al. elaborated temperature sensors in detail and provided a survey of state-of-the art examples. Chapter "Impact of AI and Machine Learning in Smart Sensor Networks for Health Care" presents a detailed study of machine learning approaches utilized on wireless smart sensor networks (WSSN) in healthcare systems. S. KajaMohideen et al. have also touched upon new scientific accomplishments of AI in WSSN for healthcare application. Chapter "ML Algorithms for Smart Sensor Networks" continues with the review of machine learning algorithms employed in SSNs. Author Geetika focused on design, operational and non-operational issues and presented detailed survey to cover important related aspects of machine learning in smart sensor networks. Chapter "Energy Efficient Smart Lighting System for Rooms" presents energy-efficient smart lighting system for rooms. In this chapter, author Madhavi et al. focused on techniques, tools, related work of existing lighting systems and comparative approaches for the implementation mechanism for smart lighting system. In today's scenario, 5G networking requires faster software automation in the existing environment. Chapter "QUIC Protocol Based Monitoring Probes for Network Devices Monitor and Alerts" formulates an approach for enhancing the HTTP-based monitoring without affecting the current services. Authors have also mentioned that the proposed approach is beneficial for network and information technology infrastructure-related technologies onshore as well offshore in collocated presence over different IP addresses over public, private and hybrid cloud architectures and monitoring. Handshake challenges for inside conflict environment are also discussed in Chapter "QUIC Protocol Based Monitoring Probes for Network Devices Monitor and Alerts." Further, there is a need to maintain and establish a secure environment in smart sensor networks. Chapter "External Threat Detection in Smart Sensor Networks Using Machine Learning Approach" presents the detection of threats occurring in IoT applications and classification of attack types in detail for an in-depth understanding of the attacks. This chapter also inculcated machine learning approaches for external threat detection in smart sensor networks.

Chapter "Towards Smart Farming Through Machine Learning-Based Automatic Irrigation Planning" introduces smart farming and irrigation scheduling and proposes a framework for smart farming through machine learning-based automatic irrigation. This chapter begins with a cleaning of the data set to effectively predict water needs. The process of data extraction is based on combined tool for data mining and knowledge discovery on irrigation and water needs. Further, authors Asmae et al. discussed a case study for water requirements of grain corn. Finally, Chapter "Graph Powered Machine Learning in Smart Sensor Networks" presents basic principles to understand graph models as a useful tool to produce better outcomes to solve complex real-world platforms. This chapter covers the framework-based graphical features with some deep learning approaches such as graph convolutional network (GCN) approach, deep graph convolutional neural network (DGCNN) and window-based approach with graphical features.

Authors have chosen a selection of specific topics related to smart sensor networks: impact of AI and ML learning in smart sensor networks, related AI-based technologies directed at smart sensors and devices from a modern perspective, efficient ML algorithms, graph-powered machine learning concepts and algorithms, security concerns, energy conservation, monitoring of network devices and incorporation of machine learning algorithms in social and intelligent real-life applications. Each chapter provides a rationale for the topic being covered as per the scope of the book and fundamental details where required.

This book is primarily targeted at postgraduate students, researchers, academicians and all those interested in exploring advances in computing and its next-generation applications. This book will be helpful to IT practitioners and executives. This book will focus on the multifaceted development of new subsequent evolutionary development and their implications and outline direction for further research.

Ghaziabad, India	Umang Singh
Auburn, USA	Ajith Abraham
Vilnius, Lithuania	Arturas Kaklauskas
Kaohsiung, Taiwan	Tzung-Pei Hong

Contents

About the Editors

Dr. Umang Singh is Associate Professor at the Institute of Technology and Science, Ghaziabad, Uttar Pradesh, India.

Dr. Umang received Ph.D. degree in Computer Applications, University School of ICT, GGSIPU, Delhi, India. She has been into research and academics for more than 18 years. She is renowned for her keen interest in the area of mobile networks, WSN, VANETs, IoT, predictive analytics and machine learning. She has published over 80 research papers in reputed journals and conferences including ACM, Elsevier, Inderscience, IEEE, Springer indexed in SCI/ESCI/SCIE and Scopus. She served as Guest Editor for special issues of journals which include *International Journal of e-Collaborations* (IGI Global, USA, 2020), *Soft Computing Techniques and Applications, American Journal of Artificial Intelligence*, SciencePG, NY, USA, 2020 and *International Journal of Information Technology*(BJIT 2010), authored 2 books and edited six conference proceedings, 3 souvenirs and 2 books. She is Member of Editorial Board of reputed journals including Inderscience IJFSE (Switzerland), Wroclaw University, Poland, Board of Referees for *International Journal of Information Technology*, BJIT, and Springer journals, and Technical Program Committee Member of national and international Conferences. She has been delivering invited talks, guest talks at prominent places and organizations including Indian Air Force. She has received Certificate of Appreciation from VSM, Airforce Station Hindan, Ghaziabad, Uttar Pradesh. She has received "Young Active Member award" for year 2007–2008 from Computer Society of India, Delhi, India. She has also received "Young Faculty in Science" in Year 2017 from VIFA, Chennai, India. She is Senior Member of IEEE and Life Member of Computer Society of India (CSI), M-IAENG and M-WIE.

Dr. Ajith Abraham is Director of Machine Intelligence Research Labs (MIR Labs), a Not-for-Profit Scientific Network for Innovation and Research Excellence connecting Industry and Academia. As Investigator/Co-Investigator, he has won research grants worth over 100+ Million US$ from Australia, USA, EU, Italy, Czech Republic, France, Malaysia and China. He works in a multi-disciplinary environment involving machine intelligence, cyber-physical systems, Internet of things, network security, sensor networks, Web intelligence, Web services, data mining and applied to various real-world problems. In these areas, he has authored/co-authored more than 1300+ research publications, out of which there are 100+ books covering various aspects of computer science. One of his books was translated to Japanese, and few other articles were translated to Russian and Chinese. About 1000+ publications are indexed by Scopus, and over 800 are indexed by Thomson ISI Web of Science. He has more than 37,000+ academic citations (h-index of 91 as per Google Scholar). He has given more than 100 plenary lectures and conference tutorials (in 20+ countries). Since 2008, he is Chair of IEEE Systems Man and Cybernetics Society Technical Committee on Soft Computing (which has over 200+ members) and served as Distinguished Lecturer of IEEE Computer Society representing Europe (2011–2013). Currently, he is Editor-in-Chief of *Engineering Applications of Artificial Intelligence*(EAAI) and serves/served the editorial board of over 15 International Journals indexed by Thomson ISI. Dr. Abraham received Ph.D. degree in Computer Science from Monash University, Melbourne, Australia (2001), and Master of Science Degree from Nanyang Technological University, Singapore (1998).

Prof. Dr. Arturas Kaklauskas is Professor at Vilnius Gediminas Technical University, in Lithuania; Head of the Department of Construction Management and Property; Laureate of the Lithuanian Science Prize; Member of the Lithuanian Academy of Sciences; Editor of *Engineering Applications of Artificial Intelligence*; International Journal and Associate Editor of journal *Ecological Indicators*. He contributed to nine Framework and five Horizon 2020 programs projects and participated in over 30 other projects in the EU, USA, Africa and Asia. The Belarusian State Technological University (Minsk, Belarus) awarded him Honorary Doctorate in 2014. His publications include nine books and 143 papers in *Web of Science Journals*. Fifteen Ph.D. students successfully defended their theses under his supervision. The Web of Science H-Index of Prof. A. Kaklauskas is 27. Web of Science Journals have cited him 2471 times and average citations per article—17.28. His areas of interest include affective computing, intelligent tutoring systems, affective intelligent tutoring systems, massive open online courses (MOOCS); affective Internet of things; smart built environment; intelligent event prediction, opinion mining, intelligent decision support systems, life cycle analyses of built environments, energy, climate change, resilience management, healthy houses, sustainable built environments, big data and text analytics, intelligent library, Internet of things, etc.

Prof. Tzung-Pei Hong received his B.S. degree in chemical engineering from National Taiwan University in 1985 and his Ph.D. degree in computer science and information engineering from National Chiao-Tung University in 1992. He served at the Department of Computer Science in Chung-Hua Polytechnic Institute from 1992 to 1994 and at the Department of Information Management in I-Shou University from 1994 to 2001. He was in charge of the whole computerization and library planning for the National University of Kaohsiung in Preparation from 1997 to 2000 and served as the first Director of the library and computer center in National University of Kaohsiung from 2000 to 2001, as Dean of Academic Affairs from 2003 to 2006, as Administrative Vice President from 2007 to 2008 and as Academic Vice President in 2010. He is currently Distinguished Psrofessor at the Department of Computer Science and Information Engineering and at the Department of Electrical Engineering, National University of Kaohsiung, and Joint Professor at the Department of Computer Science and Engineering, National Sun Yat-sen University, Taiwan. He got the first national flexible wage award from the Ministry of Education in Taiwan.

He has published more than 500 research papers in international/national journals and conferences and has planned more than fifty information systems. He is also Board Member of more than forty journals and Program Committee Member of more than five hundred conferences. His current research interests include knowledge engineering, data mining, soft computing, management information systems and WWW applications.

Smart Sensors and Devices in Artificial Intelligence

An Overview of Artificial Intelligence Technology Directed at Smart Sensors and Devices from a Modern Perspective

Ana Carolina Borges Monteiro, Reinaldo Padilha França, Rangel Arthur, and Yuzo Iano

Abstract This chapter aims to provide an updated overview consisting of a scientific major contribution regarding Artificial Intelligence (AI) directed at sensors, in applications involving intelligent technologies such as Machine Learning or Deep Learning. And so, approaching and addressing the topic with a bibliographic background, synthesizing the potential of technology. In this sense, understanding what is AI is essential since it is one of the main disruptive technologies of today, which is a branch of computer science that focuses on building computers and machines capable of simulating intelligent behavior. AI is the ability of electronic devices to function in a way that resembles human thinking, evaluating that the technology involves a grouping of various technologies, such as algorithms, learning systems, neural networks, among others that can simulate capacities related to intelligence, such as reasoning, perception of the environment and the analysis ability for decision making, which can be aimed at the industry in processes automation taking advantage of intelligent machine learning. This implies perceiving variables, making decisions, and solving problems, i.e., operating in a logic that refers to reasoning. The sensor is a device sensitive to a certain quantity, which changes its state in proportion to the stimulus on it, i.e., the magnitude of the change of state leads to the knowledge of the magnitude that caused it. In this perspective, it can be programmed and exchange data with various elements of the communication network that they are part of, including even other sensors, enable a variety of integrated applications.

A. C. B. Monteiro (✉) · R. P. França (✉) · Y. Iano
School of Electrical and Computer Engineering (FEEC), University of Campinas—UNICAMP, Av. Albert Einstein—400, Barão Geraldo, Campinas, São Paulo, Brazil
e-mail: monteiro@decom.fee.unicamp.br

R. P. França
e-mail: padilha@decom.fee.unicamp.br

Y. Iano
e-mail: yuzo@decom.fee.unicamp.br

R. Arthur
Faculty of Technology (FT), University of Campinas—UNICAMP, Paschoal Marmo Street, 1888—Jardim Nova Italia, Limeira, Brazil
e-mail: rangel@ft.unicamp.br

© The Author(s), under exclusive license to Springer Nature Switzerland AG 2022
U. Singh et al. (eds.), *Smart Sensor Networks*, Studies in Big Data 92,
https://doi.org/10.1007/978-3-030-77214-7_1

Keywords Artificial intelligence · Machine learning · Smart sensors · Smart devices · Technologies

1 Introduction

Smart sensors are fundamental parts of factories that adhere to the principles of Industry 4.0. After all, they are the ones that allow the equipment to monitor its activities and, thus, work autonomously and preventively. The sensor is a device sensitive to a certain quantity, which changes its state in proportion to the stimulus on it. The magnitude of the change of state leads to the knowledge of the magnitude of the greatness that caused it [1].

There are several types of sensors, temperature sensor, level sensor, ultrasonic sensor, among others, taking into account industrial applications, the information produced by them is communicated to a receiver in an electronic way to then generate decision making. However, until recently, sensor applications were normally limited to producing just one type of information, which resulted in specific decisions. Today, however, mainly with the development of efficient and reliable communication network protocols, such as the Internet itself, it is possible to integrate several sensors and their receivers into a network [2].

Or even artificial intelligence has been gaining more and more space in different areas of people's daily lives, and health is one of them. The new utility for the technology promises to help reduce heart failure cases by predicting the problem days in advance. It is a sticker equipped with a sensor that monitors heart rate, sleep quality, posture, physical activity, and breathing patterns, sending the data to an application [3, 4].

With this, intelligent sensors allow the development of systems that optimize, update, and reconfigure tasks. In addition, their performance can be monitored and modified remotely. These sensors that can be reprogrammed and exchange data with various elements of the communication network that are part (including other sensors), enabling a variety of integrated applications, are called intelligent sensors [5].

An Artificial Intelligence solution can be used to monitor the company's machines and computer systems. In the case of equipment, it can use data from sensors, cameras, records in monitoring software, among other sources. Regarding the systems, their information comes from databases, reports, histories, among others. And with respect to the data obtained, the technology has the ability to discover bottlenecks, failures, and other weaknesses in the company's processes, reducing errors and increasing operational efficiency. This reduces costs and avoids difficulties for the teams [3, 4, 6].

Or even considering the performance of intelligent sensors that go from underground to air and the analysis of data sets feed the artificial intelligence that gives fluidity to its streets and helps ensure the well-being of its inhabitants. With a focus on urban intelligence, technologies boost new solutions to problems that have become like floods, violence, congestion, and pollution [7].

Therefore, this chapter aims to provide an updated overview of artificial intelligence because of the joint application horizon with sensors and smart devices, addressing examples of application potential in the industry, approaching thematic from a bibliographic background, synthesizing the potential of technology.

It is worth mentioning that this manuscript differs from the existing surveys since a "survey" is often used in science to describe and explains the theory, documenting how each discovery added to the store of knowledge, talking about the theoretical aspects, how the academics piece fits into a theoretical model. While this overview (scientific major contribution) is a scientific collection around the topic addressed, relating that this type of study is scarce in the literature, offering a new perspective on an element missing in the literature, dealing with an updated discussion of technological approaches, exemplifying with the most recent research, applications, techniques, and tools focused on the thematic, summarizing the main applications today.

2 Methodology

This research was developed endorsed on the analysis of scientific papers and scientific journal sources referring to Artificial Intelligence towards Smart Sensor and Devices, concerning evolution and fundamental concepts of technology aiming to gather pertinent information regarding thematic. Thus, also enabling to boost more academic research through the background provided through this study.

In this sense, this survey is carried out on the bibliographic inspection of the main research of scientific papers related to the thematic of Artificial Intelligence, published in renowned bases in the last 5 years, such as Web of Science, Scopus, Google Scholar, EI-Compendex, SciELO, Springerlink, IEEE Xplore and more.

The differential of this paper is to address the studied theme focusing on its use in modern perspectives derived from intelligent sensors and devices, based on current examples of the applicability.

3 Artificial Intelligence Concept

Artificial Intelligence is the ability of electronic devices to function in a way that resembles human thinking, evaluating that the technology involves a grouping of various technologies, such as algorithms, learning systems, neural networks, among others that can simulate human capacities related to intelligence, such as reasoning, perception of the environment and the analysis ability for decision making. This implies perceiving variables, making decisions, and solving problems, i.e., operating in a logic that refers to reasoning [8, 9].

Thus, Artificial Intelligence is the study and construction of "intelligent computational systems", which have the cybernetic capacity to think and act rationally, composed, and modeled as intelligent agents. With regard to the Intelligent Agent, this perceives its environment through sensors and acts on that environment through actuators. For better visualization, this concept can be applied to a robotic agent, which has sensors related to cameras, microphone, keyboard, an infrared detector, among others, and actuators related to the various types of motors that this may have, or even a claw of a robot, video, call to a program, writing to files, speaker, printer, among others. This can be related that while the hardware is the physical part of a machine, the software is the logical part, i.e., the digital "brain". Thus, an AI agent is any entity that perceives its environment through sensors, that is, robots, web crawler, autonomous vehicles, Siri, industrial controller, and acts on the environment through actuators, i.e., that there is a relationship between the environment and agent (physical environment/robots; or even software environment/softbots) [9, 10].

Evaluating an Artificial Intelligence system, it can develop processes that involve regressions, correlations, structuring analyzes of the data generated, among other aspects, which serve as a basis for managers to make decisions. Even more, if it is mainly linked to a Big Data solution, which is capable of handling a huge volume of unstructured data. Still reflecting that AI contributes to the automation of logical, analytical, and cognitive activities, generating greater speed in the treatment of information, serving as a complement to the automation of physical tasks, especially in industrial production, provided by robotic machines [11].

Regarding the structure of intelligent agents, that is, the rational agent which is the central concept in AI, and the design of the agent program, i.e., the function that implements the mapping between perception and action. Still analyzing that the agent program is executed in a computational device architecture including sensors and actuators, still pondering on the Agent specification components, PEAS (Performance, Environment, Actuators, Sensors) [12].

The concept of AI (Fig. 1) is related to the ability of technological solutions to perform activities in a way considered intelligent, which can "learn for themselves" due to learning systems that analyze large volumes of data, enabling them to expand their knowledge. AI is also a field of science, whose purpose is the study, development, and use of machines to carry out human activities in an autonomous way, linked to Machine Learning, voice recognition, and vision, among other technologies [13].

AI has the properties to assist in the simplification of analysis processes, especially those that value data-driven decision making, since it is capable of organizing and providing greater clarity to "cloudy" or "confusing" data", Which leads to the establishment of digital strategies. Finally, AI is developed so that devices created by man can perform certain functions without human interference [14, 15].

Still reflecting on the Knowledge Representation which aims to reduce problems of intelligent action for search problems. Considering that a good representation makes important objects and relationships explicit, exposing the internal restrictions inherent to the modeling problem by AI [16, 17].

Fig. 1 Artificial intelligence
concept illustration

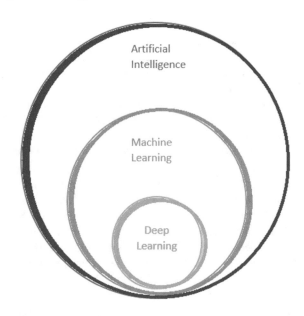

One of the main facts about joint intelligence is that it is a technology capable of providing cost reduction, production optimization, and activity management, generating competitive advantages for businesses, which makes the technology a strategic solution. Still pondering the great advantage of AI is that it is allowed to perform tasks much faster, and with a degree of accuracy thousands of times greater than that of humans [16, 17].

3.1 Machine Learning

Machine Learning involves a method of data assessment that automates the development of analytical standards, designing programs that learn to make predictions based on data alone, without the assistance of a human programmer. The technology is based on the conception that technological systems can learn using data, in order to discover patterns, make decisions, and improve with little human interference, improving the performance of activity over time [18].

Still reflecting that people do not always notice how IoT (Internet of Things) can be useful, since they are used to recognizing the world through human senses, however, IoT technology captures the characteristics of a person to be learned, capturing all their actions that converge on a routine and well-being, without even someone having to say or push buttons [19].

Explicitly about this context come from devices with sensors capable of using this scenario through an accelerometer sensor, gyroscope sensor, compass sensor, microphone, GPS, camera, light sensor, among others, it is possible to build an application that knows the exact time when a person wakes up according to the days of the week, to know the path this individual takes, the time he leaves and returns from work/college. And from the acquisition of these data, a house can be automated specifically for that individual or his family, through Machine Learning. This also makes it possible to control and read sensors even when not present in the same environment, uniting the traditional IoT to AI, based on the step beyond capturing the performance of sensors, is the way to use this data for a means and an end, making the user's life better [20, 21].

Still reflecting on Logistic Regression is the most used statistical method to model categorical variables, this technique is like a linear regression analog for classification problems. Which appears when you want to categorize a variable by class. Considering that the technique will provide a forecast between 0 and 1, so that it is possible to interpret its results as a valid probability [21].

The machine learning field has algorithms used in applications such as music recommendations, spam filters, and fraud detection, among many others with the ability to take advantage of the effectiveness of a situation and use it in the practice of another activity. Technology has the capacity to resolve new situations quickly and successfully, adapting to them through acquired knowledge. Thus, Machine Learning technology is capable of providing computational capacity, as well as data, algorithms, APIs, among other solutions for the design, training, and application of models on machines, devices, applications, processes, among others [22].

Pattern recognition is a part of Machine Learning that focuses on regularities in a given data scenario, and can be of a supervised type, i.e., the context in which the algorithm has already been fed with patterns that it should look for, or even unsupervised, i.e., for those contexts in which the algorithm discovers new patterns [23–25].

Among the types of sensors used for this technology:

- Gas sensors are used to monitor changes in air quality and detect the presence of various gases.
- Chemical sensors are present in industries detecting the presence of toxic or combustible gases, as well as gases from coal mines, the petrochemical industry, and some types of manufacturing, such as rubber, paint, plastic, and pharmaceutical. Highlighting sensors detecting carbon dioxide, methane gas, carbon monoxide, nitrogen oxide, oxygen, and ozone.
- Level sensors are used to determine the level or quantity of fluids, liquids, or other substances that flow in an open or closed system, especially applicable for measuring fuel levels. In agribusiness, ensuring that work processes are always continuous, avoiding unnecessary stops, has properties to measure the operating time of tractors and other agricultural machines, advising at the right time when they need to be replenished. This is still applied in the measurement of fuel and other liquids in open or closed containers, monitoring of tides, and water reservoirs.

3.2 Computer Vision

The purpose of computer vision is to help computers identify and process images in the same way that human beings do, through learning and ultimately distinguishing the faces of different people. Computer vision teaches machines to recognize the different objects that are "seen" through the sensors of a camera on a given device [26].

Computer Vision is fully connected with Machine Learning, which is the area of science that develops theories and methods aimed at the automatic extraction of useful information contained in images, in order to create and transmit this information to machines in an understandable way. Technology analyzing individual pixels, identifying different colors, or even converting them to numerical values and then looking for patterns to identify groups of pixels with similar colors and textures, helps machines to identify different objects [26, 27].

Among the types of sensors used for this technology:

- Passive Infrared used for home security by detecting body heat (infrared energy)
- Ultrasonic sending pulses of ultrasonic waves and measuring the reflection of a moving object, tracking the speed of sound waves.
- Microwaves sending pulses of radio waves and measuring the reflection of a moving object, covering an area larger than the infrared and ultrasonic sensors, however, vulnerable to electrical interference.
- The proximity sensor detects the presence or absence of a nearby object and converts it into a signal that can be easily read by the AI system or by a simple electronic instrument. Proximity sensors are widely used in the retail sector, as they detect customer movement in one location, or even used for parking availability in places such as shopping malls, stadiums, and airports.

3.3 Deep Learning

Deep Learning is a different type of Machine Learning which involves neural networks with several layers of abstraction, which allow data mining and pattern recognition in large data sets that would not otherwise be quantified or classified promptly, and classifying applications supported by data sets. A neural network can solve very complex problems because there is an enormous number of neurons working together [28].

This allows the predictive algorithms to work directly with the information that previously required the action of human beings to separate and classify, evaluating that this type of learning process occurs between their layers of mathematical neurons, in which information is transmitted through each layer, since in this structure, the output of the previous layer is the input of the posterior layer. Deep learning is built on neural networks in such a way that it looks like neurons in a human brain, i.e., in a neural network, artificial neurons are organized in inter-connected layers. Provided an input layer to receive external data and an output layer that dictates how the system will respond to the information. Among these layers, there are other additional "hidden" layers of neurons, such as CNN (Convolutional Neural Network) that process data by giving a numerical weight to the information they receive from the previous layer, and passing that information to the next layer of the network [29].

Reflecting on Artificial Intelligence, in general technology is divided into a symbolic approach, in which the mechanisms effect transformations using symbols, letters, numbers, or words. And so, they simulate the logical reasoning behind the languages with which human beings communicate with each other. And also, concerning the connectionist approach, since it is inspired by the functioning of human neurons, simulating the mechanisms of a person's brain, and in this context is Deep Learning, in which the ability of a machine to acquire deep learning, imitating the neural network of the brain [30].

Still reflecting on the evolutionary approach, which uses algorithms inspired by natural evolution, concerning metaheuristics, that is, the simulation of concepts such as environment, phenotype, genotype, perpetuation, selection, and death in artificial environments. Or even neural networks, which also fall into this classifi-cation, that is, systems that require thousands and thousands of data for models to satisfactorily perform their complex registration tasks. And in some cases, millions of these data are needed to perform at the same level as a human being [31].

Still reflecting on a multilayer perceptron is a neural network with one or more hidden layers with an undetermined number of neurons, but with more than one layer of neurons in direct feed, being able to relate knowledge to several output neurons. Such a network is composed of layers of neurons linked together by synapses with weight, even considering that the hidden layer is not possible to predict the desired output in the intermediate layers [32, 33].

Thus, technology "trains" machines to perform activities as if they were human, whether in image identification and speech recognition, or even data processing. The technology is applied directly to companies, has the properties of automated decision and interaction processes, making this high level of automation reduce costs and risks and even allow for greater revenue through better marketing and digital sales segmentation [34, 35].

Among the types of sensors used for this technology:

- Inductive sensors are used for non-contact detection by discovering the presence of metallic objects through an electromagnetic field or a beam of electromag-netic radiation.
- Photoelectric sensors are composed of light-sensitive parts responsible for detecting the presence or absence of an object (ideal alternative for inductive sensors), which can also be used for long-distance detection.
- Ultrasonic sensors are used to detect the presence or measure the distance of targets, similar to radar or sonar.

3.4 Natural Language Processing (NLP)

Natural Language Processing is related to the study and the attempt to reproduce development processes linked to the functioning of human language, employing solutions based on Deep Learning making computers understand, process, and manipulate human language. As far as the need for a machine is able to "understand" a huge amount of information, from grammatical rules and syntax to colloquialism and even accent [36].

Analyzing a voice recognition system, in which the human voice becomes audio data, which is then converted to text data in another complex process. This text data can then be used by a "smart" system in a number of applications such as translators or to control devices such as televisions, aimed at Smart House applications [37].

Or even through Machine Learning contained in a corporate identity verification system or tracking and filming of a proprietary drone or an object, it can improve natural language interfaces, allowing smart speakers to react more quickly, interpreting voice instructions locally, executing basic commands, such as turning lights on/off, or adjusting thermostat settings, even if Internet connectivity fails [38].

Through Natural Language Processing, the machines can understand texts, which derives the context recognition, information extraction, development of summaries among other aspects, or even making it possible to compose texts starting from data obtained by sensors capturing audio in devices scattered in the environment. The technology can be used in areas such as customer service or even in the production of corporate reports [39].

3.5 Big Data

Predictive analysis is the ability to identify the probability of future results based on data, statistical algorithms, and Machine Learning techniques, around Big Data, related to what is about large data sets that need to be processed and stored [40].

Based on big data, therefore, there are programs capable of doing this type of analysis, identifying trends, predicting behaviors, and helping to better understand users' current and future needs. And even qualify the decision-making in various machines, equipment, and software, taking artificial intelligence to a new level, based on data collected by sensors. In this sense, all sensors generate and collect data, whether the types of sensors such as Infrared, Ultrasonic, Microwave, Proximity sensor, Inductive sensors, Photoelectric sensors, Gas sensor, Chemical sensors, or even Level sensors, or other types, they need Big Data technology to process all that volume. While for AI to be implemented and adapt well to the installed structure, it needs a lot of data, which is also important for the process of integrating the smart platform with workflows and software, as well as for increasing the accuracy of the information generated by it [41, 42].

In this context, a Big Data Analytics solution is important to generate insights and information from unstructured data, which will be used by an Artificial Intelligence solution, using IoT and sensors to capture information from the equipment that feeds it, grouping the generated data for the solutions [43].

4 Smart Sensor and Devices Concepts

The sensor is a device sensitive to a certain quantity, which changes its state in proportion to the stimulus on it, i.e., the magnitude of the change of state leads to the knowledge of the magnitude that caused it. In this perspective, intelligent sensors are fundamental parts of the factories that adhere to the principles of Industry 4.0, after all, they are the ones that allow the equipment to monitor their activities and, thus, work with autonomy and in a preventive manner. Or more simply, a simple example of a sensor consists of the home thermometer (temperature sensor), which consists of a fluid that expands according to the temperature. Thus, the magnitude of the expansion is associated with the temperature value around the thermometer [44, 45].

There are several types of sensors, more specifically in industrial applications, the information produced by them is communicated to a receiver electronically to then generate decision making, through intelligent technologies such as Machine Learning or Deep Learning, strands of AI [45].

Sensor applications, in general, tend to be limited to producing just one type of information, which leads to specific decisions. Currently, mainly with the development of efficient and reliable communication network protocols, such as the Internet, it is possible to integrate several sensors and their receivers into a network. This allows the development of systems that self-optimize, update, and reconfigure tasks, through digital learning technologies, still evaluating the characteristics in which the performance can be monitored and modified at a distance [46].

The sensors that can be reprogrammed and exchange data with various elements of the communication network that they are part of, including even other sensors, enable a variety of integrated applications, which are called intelligent sensors. Still pondering the current panorama of the relevant data collection and analysis standards and protocols, consisting of heterogeneity, i.e., different technologies, it is necessary to connect all the data and translate them into a single language. By transforming all this information into a single language, decision making becomes better, an Open and accessible protocol for representing data that allows the dynamic exchange of information between sensors [47].

Over the past few years, machines have acquired the ability to generate a large volume of data, using sensors (soil sensors, climate sensors, sensors capturing images, sensors connected in real-time operating autonomously and having their movements controlled by GPS, and even precision sensors, among others), from those georeferenced from agricultural and environmental operations, such as equipment speed, engine temperature, fuel consumption, among others information, even technologies that move towards autonomous industrial machines that work in the future without operators [48].

The advantages generated for the industrial manager come through intelligent sensors (sensors for measurements of various magnitudes, such as pressure, humidity, temperature, position, among others), which check the processes and make alerts in case of any irregularity that could put employees in danger, thus reducing the risks of losses and unwanted results. And in this respect, they bring a series of advantages to the factories that use it, ranging from more adequate monitoring of processes to more agility in decision making. In addition to the benefits of improving productivity, with the use of technology through automation, or even with the elimination of waste and rework, or even with robotization through greater safety for employees. In this sense, with automation, it is possible to monitor the entire industrial complex 24 h a day, 7 days a week [48, 49].

Evaluating the context of a vehicle manufacturer, this has the capacity to produce models with individualized characteristics according to the wishes of each customer, without compromising the speed and quality of production, due to the communication of all these elements in the same network, from the line from production to resale, including logistics [48, 49].

In this context, the Internet of Things (IoT) consists of the possibility of physical objects having the characteristics to communicate with each other, via the internet. Through interconnected sensors that allow the identification of each device on the network. Still reflecting on the future perspective that a printer may order, or even notify an intelligent system regarding the purchase of print cartridges directly from the supplier when the ink levels are below a certain limit [50].

Intelligent sensors can still be added to Artificial Intelligence within Industry 4.0, in the sense of spell checkers and search pattern recognition algorithms, which store user data and already offer word corrections increasingly consistent with the meaning of the sentence, or suggestions for purchases based on recent research [50].

4.1 Sensor Applications in Industry 4.0

The applications of intelligent sensors (measuring various magnitudes, such as pressure, humidity, temperature, position, among others) in Industry 4.0 are numerous, in terms of depending on the needs of each industry, a variant of its niche. Assessing the context of the shop floor, the sensors can be found on the equipment, on the walls, windows, on the clothes, or even on the PPE (personal protective equipment) of the employees, on the vehicles among many others [51].

The presence of sensors (relating those with properties to collect temperature and humidity in a single sensor with digital output for communication. Or still pondering that in general, this intelligent sensor can combine (pressure, humidity, temperature, position, among other magnitudes) optimizing the space in equipment and facilitates the engineering work) on machines and devices increasingly connected to the internet allows the collection of a huge amount of data, which can be used to detect production failures, predict production patterns, or even understand the consumption profile of customers. Pondering on the analysis tools, such as Big Data that allows optimizing the factory's process, identifying trends, and even developing customized products [52].

Still discussing with respect to the connectivity of the machines to the internet in which all the data generated is transmitted via the internet to cloud technologies, forming a large database of operations and the industrial environment, creating big data. And so, making it possible to interfere, through AI or strands on the generated data analyzed, providing more efficient operations, at the moment it is being carried out, in cases of deviation from the plan. With respect to the information generated by the sensors, they can be analyzed, allowing them to identify patterns and events with more agility and assertiveness. What facilitates predictive maintenance, for example, allowing immediate recovery before a failure, thereby avoiding corrective maintenance and eventual losses associated with failure and even stoppages of production lines [53].

The lack of preventive maintenance increases the risk of accidents, since in general there is little investment in preventive maintenance of machinery and equipment by companies, which routinely act only when a breakdown or a defect in the machinery occurs, reducing the productivity of the entire operation. And so, it exposes the worker to situations that compromise his physical integrity, when trying to solve the problem. What through the AI fed by the data collected by several sensors, has properties and characteristics to develop a routine of regular maintenance, guaranteeing a greater safety of the workers, a reduction in the cost of repairs, and an increase in productivity [51–53].

Since Artificial Intelligence-based programs are independent and based on pattern recognition, that is, analysis of behavior and habits that allow the program and users to be in tune, which can help to make applications and solutions through automation and robotics manage the production of a factory. Determining the time required for each part or group of parts in the manufacturing process, considering temperature, pressure, the humidity of the day, in addition to the quality of the raw material. Still allowing the most intelligent and optimized use of time, which allows the desired quality with reduced time and repair costs [54].

Finally, the greater use of image processing techniques, which can be captured by mobile devices such as drones, or sensors, provide a computer vision, when the machine itself has mechanisms to see patterns, events, and objects. And so, making the images and data captured through sensors be processed in real-time and the machines respond immediately to ongoing operations, when desired and triggered [47, 55].

Currently, the limit of the useful application of intelligent sensors is more in creativity than in the technology itself, which to be activated, basically depends on good equipment, good programming, and an efficient data transit network, such as the internet. As a result, they become a vital component, whether in Industry 4.0 or another field of action, and have enormous power to generate improvements and increase productivity, through data collection, suggesting new dynamics for the applied sector [48, 55].

5 AI Application on the Smart Sensor, Analytics, Sharing, and Control Perspective

Due to the capabilities of AI Technologies (Fig. 2) in relation to data analysis and processing, they are very useful in the evaluation of indicators and in supporting decision making in the Financial sector of an organization, as can provide suggestions for measures to be applied in the business. Acting directly in management processes, such as automating price calculation or even assessing which application can provide the best Return on Investment (ROI) [56].

An Artificial Intelligence solution can be used to monitor a company's machines and computer systems, considering that the equipment relies on the use of data from sensors (motion sensors, presence sensors, distance sensors, color sensors, position sensors), cameras, records in software monitoring, among other sources. This means that it has the ability to discover bottlenecks, failures, and other weaknesses in the processes, reducing errors, and increasing operational efficiency [8–10].

Reducing costs and avoiding difficulties in relation to systems that use information from databases, reports, histories, among others. Still considering the properties of making predictions about possible difficulties, allowing the company to take action on time, generating diagnoses of different sectors and processes, contributing to the mapping of risks, and also opportunities of the company. Still allowing the possibility of checking with a relative frequency of key business performance indicators, or even evaluating the analysis processes that usually present a low cost, contributing to a higher level of accuracy in the defined strategies, based on the data collected by sensors and processed by AI technologies [8–10].

Furthermore, even in the Financial sector, AI has a prominent role in preventing fraud, since it has properties to automate and optimize the analysis of credit and risk granting, or even insurance. Assessing which entities in this sector through AI can reconcile bulk transactions, i.e., data volume. Through the data obtained by interacting with customers, AI tools are able to provide an expanded understanding of their demands, expectations, and desires [8–10].

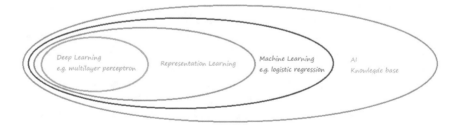

Fig. 2 AI technologies illustration

In an organization's Human Resources (HR) department, AI technology in the face of its ability to perform operations that require less analysis and reasoning, can contribute to redirecting employees/employees in repetitive tasks to higher added value activities [57].

Or even pondering about the application of the Machine Learning component in customer service, through chatbots and systems with intelligent natural language processing replacing human attendants, still considering the characteristic of technological disposition 24 hours a day. With online retail support, online store algorithms, using AI technologies, have properties to recognize users' shopping patterns while they can present them with offers according to their preferences [57].

The AI will be able to collaborate in the mapping of trends, acting in the Marketing sector, in relation to behaviors and opportunities with the public, in addition to foreseeing demands, given the potential to help in serving the public in different aspects, as in the delivery of more communication efficient, increased transaction agility or even extended personalization. Considering the application of chatbots, as they can interact with consumers, but have lower capacity compared to robust AI solutions. Which can be used in message boxes of websites and social networks, virtual service rooms, instant messengers, among many others [58].

Another activity that AI improves is the evaluation of user behavior, so that they are able to better answer consumer questions based on analyzes made in the content of the dialogues. This is because it is able to analyze digital content and apply profile segmentation algorithms, based on the habits of the public, which thereby recommend products that consumers are most likely to buy [58, 59].

The retargeting process that derives from impacting the consumer more than once, that is, means aiming again at a target that has already been pointed out. Within the reality of marketing, it means impacting the target audience again, it is another process favored by AI, as long as it is possible to identify users' online shopping and browsing habits. And so, AI strands help to automate this strategy, sending an offer or an alert about the price reduction of an item to those more likely users who almost bought it before, but abandoned the purchase process, for example [59].

Still reflecting on the many accessories that people use in their daily lives integrating IoT technology, ranging from some household appliances to accessories that collect and send information, up to door sensors (presence sensors, distance sensors, position sensors) that control children's arrival and departure times after school. And in this perspective, digital solutions involving the union of IoT and AI through systems dedicated to security perform different functions, involving different types of protection, such as patrimonial and intellectual, or even the control of work safety for the allocated professionals in companies [60].

AI is one of the technologies that spearhead the modern industrial revolution (industry 4.0), along with augmented reality, to the IoT, still mentioning the virtualization of physical elements of the factory to create copies (digital twins), performing tests in virtual simulations of the park factory in search of higher performance processes, optimized disposition of factory elements, elimination of bottlenecks, among other aspects. This results in the AI, allowing equipment to have the necessary properties to check the products without having to be operated by a human [60].

In fact, due to the IoT, AI is able to obtain data not only from virtual systems, but also from physical devices, through sensors, which are not a computer or electronic, i.e., industrial machines, relating operations carried out in production. Being able that the applications of AI in the industry are vast, since it can be used in both productive and managerial activities [60].

Assessing that while traditional data analysis is carried out using Business Intelligence (BI) techniques that monitor, gather, and organize data, analyze through AI analyze actions that occurred in the past, identifying new opportunities, and implementing strategies based on the data, adding a more prominent processing arsenal [61].

Still reflecting on AI combined with robotics, with the potential to improve manufacturing processes by mapping bottlenecks and reducing errors generated by robots. Due to the continuous monitoring of sensors, telemetry equipment, cameras, among other monitoring devices [11].

Or as for predictive maintenance with regard to analyzing data such as temperature, noise level, pressure, among others, anticipating the need for maintenance before a problem affects any equipment. Avoiding unnecessary maintenance, which can paralyze production, generating cost savings. In this context, the power of cognitive computing acts by detecting anomalies, through the analysis of huge amounts of data, collected by sensors (temperature sensors, noise/sound sensor, pressure sensor) [54].

In the logistics sector to reduce costs with real-time behavioral forecasts and guidance, AI techniques can be applied as continuous estimation adding substantial value. Managing to optimize delivery traffic routes, rationalizing the use of fuel, and shortening deadlines. Still pondering the paradigm of the smart city, which "listens and understands" its population, employing several sensors (temperature sensors, humidity sensor, sensors that send data over the internet, location sensor (like a GPS receiver)) scattered along its perimeter, adapting to dynamic parameters, such as climate and traffic, among many others. Sensors that go from underground to air and the analysis of data sets feeding AI, giving fluidity to the streets and helping to ensure the well-being of its inhabitants [47].

Still related to the security issue, through the data collected from the webcams, which through the AI, it is possible to use the laptop and desktop webcams as security and prevention devices, making it possible to insert strategies through intelligent sensors (motion sensors, presence sensor, magnetic sensor (in contrast to the presence sensor, triggering the alarm only when recording a movement, this by means of a highly sensitive magnetic detector identifies opening and break-in impacts at doors and windows, or even smoke sensor) directly on the office doors and attach the alert system to webcams. Preventing attempts at data theft, misuse of passwords by unauthorized individuals, fraud in logins, and even embezzlement of money filmed in real-time, collecting serious evidence [47].

As far as urban intelligence is concerned, AI technologies boost solutions to problems that old people have known, such as floods, violence, congestion, and pollution. Going in this direction, the use of "Big Data" can be added in the decisions of traffic departments in large cities, deriving through the collection of data on equipment and sensors (ultrasonic sensors, noise/sound sensor, optical sensor (infrared), radar sensor, magnetic sensor) spread over the urban territory, they are able to monitor and recognize the sound of shots and turmoil to alert security monitoring centers, for example [4].

As already mentioned, tools derived from AI and strands, can link with Big Data, in which an immense amount of data is produced and stored daily. And based on this abundance of information, these intelligent systems and solutions can organize, analyze, and interpret (process) the data, which are generated by multiple sources.

With regard to traffic, the routes suggested by the application generally point out the best route, coming from AI solutions aimed at interpreting data provided automatically by other users about the traffic on the roads [34].

Still relating the advent of autonomous cars that do not need a driver to guide them, they are due to a combination of various technologies and sensors (sensors of various types, capturing information in real-time about the position on the road, the relative distance of potential obstacles and other vehicles, among other aspects) that provide data for AI algorithms to guide the movement of automobiles. Some current automobiles are able to identify road signs and feature panels that inform drivers when the lights change, helping to avoid traffic violations and dangerous situations. AI-enabled sensors can be deployed to alert drivers of sudden changes that the naked eye may miss, such as an object on the road or another vehicle in a blind spot. Still pondering other communication systems from one vehicle to another, cars can alert vehicles about an accident, giving cars time to activate brakes and prevent further collisions [47].

Reflecting on the development of autonomous cars, which do not purely have the function of allowing greater comfort to people, but are the result of the idea that they can reduce the number of accidents in traffic, assuming that the computer will not make the same mistakes behind the wheel that a person commits [47].

5.1 Health Applications

Portable devices and sensors connected to an individual's body are able to send data in real-time on various aspects of their health, such as heart rate and glycemic index. Evaluating that Machine Learning and Deep Learning algorithms, through the collection of this data by these sensors, can analyze this information, managing to draw a complete picture of the patient's condition, and creating medication alerts or even detecting potential future diseases, anticipating diagnoses and recommending more assertive treatments are possible [62, 63].

Or even an accelerometer sensor that allows a patient's breathing and coughing movements to be captured, can be used on the body to monitor COVID-19 symptoms. These data are analyzed by an articular intelligence system in order to know if there is a risk of infection by COVID-19. Or even a wearable glued (equipped with sensors) to the body, monitors the user's temperature and transmits information in case of fever, aiding diagnosis, and allowing early treatment in case of flu infection or the new coronavirus [64, 65].

Still from the perspective of medical safety in hospitals, a temperature sensor can check if the patient has a fever without physical proximity, installing this type of sensor capable of measuring a person's temperature from a distance. Still combining the use of a thermal camera and facial recognition algorithms, it is possible through intelligent algorithms to detect and diagnose if a patient has a fever [64, 65].

Or even from the perspective of reducing bottlenecks in the fight against COVID-19 with respect to slowness and difficulty in processing diagnostic tests of the disease, considering the use of infrared sensors with artificial intelligence without making contact with people to quickly highlight individuals with fever, supporting screening in sophisticated ways. Thus, systems of this type can identify potentially contagious individuals, replacing a complicated manual screening process, representing initiatives using joint intelligence in the diagnosis of the disease can be applied through Big Data and Predictive Analysis to support the medical decision using and analyzing basic clinical data, such as blood tests, to present a percentage chance of the patient having COVID-19 [64, 65].

Or even through the application of RNA (Articular Neural Networks) for recurrent and evolutionary learning; with each performed and validated examination, joint intelligence can learn from digital, X-ray, or lung computed tomography images, increasing accuracy. Generating histories and analysis data for health professionals, helping the medical team to make clinical decisions, and understanding the severity of the disease, indicates. For example, if the chances of the patient having high mechanical ventilation or dying are high, provided that the information on cardiac monitors and ventilators can be collected through pressure sensors and can be integrated into these devices, and the data analyzed, allowing a deeper understanding of the patient's condition. This type of solution is suitable for the reality of emergency rooms, which receive a large volume of people and generally do not have a radiologist available [66].

6 Discussion

Understanding what is Artificial Intelligence (AI) is essential, since it is one of the main disruptive technologies of today, having the potential to considerably modify the way in which organizations operate. There is a potential for those who have adopted this technology, regardless of the area in which they find themselves, going beyond mechanical automation, encompassing cognitive processes, which generate a capacity for digital learning.

Or even reflecting on the aspect that technology has been gaining more and more space in different areas of daily life, and health is one of them, relating solutions that use AI which helps to reduce cases of heart failure by preventing the problem with days in advance. In this way, an Artificial Intelligence system is able to perform activities that are not only repetitive, numerous, and manual, but also those that require analysis and decision making.

Or even relating the smart technologies directed to health with regard to a sticker equipped with a sensor that monitors heart rate, sleep quality, posture, physical activity, and breathing patterns, sending the data to an application. With the purpose of promptly detecting changes in the patient's body, such as heart, with sufficient advance allowing doctors to initiate immediate interventions that can prevent rehospitalization, for example, or even prevent the worsening of heart failure.

The current technological tools make it possible for small and almost imperceptible sensors (motion sensors, presence sensors) to be placed in internal and external areas of the property and to send motion detections to the heads of the security center. What makes it easier to use these resources optimized with the use of IoT and enhanced through AI. Providing greater security, evaluating that when a sensor issues the alert, a team can be sent to the location for checking, if impossible, the cameras can be repositioned and checked remotely. Faced with such characteristics, places that offer restricted access to a small number of people, such as laboratories and centers with a large amount of intellectual property, benefit from this type of technological tool.

Another point of discussion is the offer of jobs, analyzing that the introduction of automation or robotization in a company will not eliminate jobs, but will replace the operational labor with a more qualified one. Assessing that a company that replaces the labor used in a manual operation, with automatic equipment or even automation, will need technicians capable of carrying out maintenance, customizations, adjustments, and the possible development of such equipment.

The advantage of greater efficiency and productivity is very evident, however, one of the disadvantages is the fact that it must result in job losses worldwide. However, not everyone will be replaced by machines, but it is a fact that the tendency is to decrease, year after year, the need for human labor in many areas. What on the other hand and perspective, new areas and activities will need new types of professionals, that is, other professions will also emerge.

An AI solution is capable of employing algorithms to perform more accurate segmentations, with a high level of process replicability, in order to suggest goods in line with the consumers' perspectives analyzed, increasing the chances of developing good commercial strategies. Assessing that the AI technologies that compose it are capable of performing the same analyzes several times, ensuring that any workflow becomes scalable. Still pondering about obtaining relevant information from reports it can also become faster, coming from text mining algorithms (text mining) capable of analyzing a document and finding information in it.

7 Geospatial AI

Still relating geospatial AI, which is the meeting of artificial intelligence with geospatial data systems, considered the frontier of satellite technology in the world, representing a revolution in the satellite sector, offering the promise of unlimited opportunities, with applications in practically all commercial and industrial areas around the world, expanding the use of geographic data far beyond what is currently done. The exploration and analysis of images and geospatial information that describe, evaluate, and visually represent physical characteristics and activities geographically referenced on Earth can be guided and defined [67, 68].

Regarding the technological aspect, there are three fundamental trends in the area of geospatial AI, with regard to the advancement of AI especially in Machine Learning; increasing the availability of geospatial data from satellites and remote sensors; and the large availability of computational capacity for high volumes of data. More precisely, Machine Learning and Deep Learning, based on neural networks, appear as promising approaches in the analysis of geospatial data, creating practical means of identifying objects and environments in very detailed images via satellite [68].

As geospatial AI technology matures, the focus should increase on automating workflows with AI, where performance and cost-efficiency issues are a priority. Reflecting on geospatial data acquisition phases, AI is focused on decisions about high-performance data archiving and data transfer rates, i.e., computational throughput. Still pondering process steps related to data preparation and model development using AI for transfers between different systems, such as general-purpose CPUs (Central Processing Unit) and specialized CPUs [68, 69].

8 Trends

The integration of AI with Blockchain and IoT with respect to autonomous cars in which the technology regulates the sensors of the car, which collects data in real-time, while the AI models act in the decision-making part of the vehicles. And in this scenario, Blockchain technology works in conjunction with AI to solve security, scalability, and trust issues and issues with the application [60].

In the same sense that cybersecurity systems with AI tend to continue to play a significant role in controlling these attacks, through the Machine Learning strand, organizations will be even better able to detect these security breaches with ease. Assessing that in general, according to cyber-attacks, it has been surpassing the existing traditional defensive measures and even increasing its heterogeneity quite quickly, in the last few years [70].

The technology will also be applied even more with respect to real-time interactions with customers, as currently, most real-time marketing activities are limited to automated responses only. With the help and momentum of AI, organizations will manage interactions with customers in real-time across all channels, developing more practices in real-time, while still being able to use AI marketing to improve customer retention. In addition to considering that AI can also help marketers to find new audiences on social media and other digital social platforms [57].

In the same sense, there will be a greater rise of virtual assistants, which are based on AI, facilitating the users' routine and automating customer service and sales tasks. More and more companies will be adopting these smart solutions to perform basic and targeted tasks. Still considering the applications of these virtual assistants in surveys that are carried out by voice, which should use PLN technology as a base [59].

Just as facial recognition is one of the main features of biometric authentication, even though there are still aspects about its imprecision, there should be a growing increase in in-depth research in this field, improving these tools more and more. Regarding facial recognition with regard to significant improvements concerning accuracy and readability [57, 67].

Still reflecting on the granularity aspect of the AI related to its constant evolution to the point of seeing smaller and smaller groups, i.e., integrating different data sets, producing insights that can be applied at a granular level, which can be tracked and scaled closely within organizations. Exemplifying the use of sales forecast in a given market, however, it is much more difficult to know the chance of a particular person to buy one of them. What requires knowledge about the individual, i.e., his habits, concerns, location, his context, among other characteristics. And this is where AI is making this possible, making it possible to zoom in on the individual. Considering that the most important thing in Big Data is not large groups, but, on the contrary, granularity, that is, acquiring the ability to see deeply each individual in a given group [60].

Patients infected with COVID-19 can be monitored by sensors in conjunction with Artificial Intelligence positioned under the mattresses of the beds. This type of system has the properties of continuously analyzing patterns in these patients, such as cardiac, respiratory, and body movements, alerting the medical team when the patient appears according to algorithmic analysis (Machine Learning or Deep Learning) is heading towards respiratory failure or sepsis [62, 63].

Still reflecting on the use of AI in health, through a sticker equipped with a sensor that monitors heart rate, sleep quality, posture, physical activity, and breathing patterns, and still sends this data to a medical application. It is possible to help reduce cases of heart failure by predicting this type of disease days in advance. Given that heart failure affects millions of people worldwide, this type of technology can reduce emergency hospitalization rates, accurately predicting the likelihood of hospitalization by allowing doctors to initiate immediate interventions that can prevent the worsening of heart failure [62, 64].

Or even reflecting on the versatility and flexibility of applied AI as a trend in use associated with patient monitoring; management and use of large amounts of data; and intelligent assistance and diagnostics. AI-monitored sensors can assess a patient's health and offer personalized care while at home. Diagnoses made and patients screened for AI can be referred to as the resources they need in less time [62, 65].

Finally, predictive medicine can become a reality with the help of AI fed through sensors, composing algorithms to predict disease risks, intervening before its onset, and anticipating a more effective treatment. By learning and leveraging data collected from patients, electronic health records, and even historical health information, AI systems will be able to maximize and optimize resources for clinical care and treatment [62, 66].

9 Conclusions

Artificial Intelligence (AI) is a branch of computer science that focuses on building computers and machines capable of simulating intelligent behavior. Still reflecting on Artificial Intelligence systems capable of doing tasks traditionally associated with human intelligence, such as voice recognition, visual perception, decision making, and language translation, among other specific tasks.

Aimed at the industry, process automation has provided benefits in the daily lives of small and medium-sized industries today. Advanced Manufacturing (Industry 4.0), through Artificial Intelligence makes use of mechanization, automation, robotization, software, among other digital technologies to innovate and produce more and better. Evaluating this scenario, in which digital technologies help industries to optimize and improve their processes, in the same sense that it generates innovation and competitive advantages.

However, it is still common, depending on the country, a good part of the segments of small industries, companies are found with production processes that are not very automated or even without any mechanization, almost artisanal. As a result, the adoption of these technologies should not be seen only as an expense, but as an investment for the business. Assessing the huge volume of data available to be decrypted on the network, but most are unstructured, requiring that in order to operate efficiently and even collect this information, advanced tools such as artificial intelligence are needed, through which the data will be visualized and worked on.

Reflecting from a future perspective, there will tend to be a period of transition between the current reality and the world in which less human work will be needed. Requiring society to deal with this through the perspective that advances in technology and innovation provide valuable insights based on data and information extracted from this work. Still being able to use this data assertively and transform it into business intelligence and resources for strategic sectors, such as security, it will be necessary to develop ways to interpret this information. The use of Artificial Intelligence can be even more fundamental to gain competitive advantages, considering that those who are successful in using this technology will be able to produce more and lower their costs, gaining an immense competitive advantage.

Through advances in technology and innovation, it is possible to extract valuable solutions from multiple sources of information, mainly by combining the analysis and extraction of valuable information from a large amount of raw data, i.e., Big Data linked to AI. Mainly related to Machine Learning, that is, the ability of the machine to continue improving its own performance without human beings having to explain exactly how to perform all the tasks assigned to it.

Still adding technological potential through perception and cognition covering a good part of this digital territory, which ranges from driving a car to predicting sales or even deciding who to hire or promote, emphasizing that AI in an increasingly short future will reach levels performance indicators in most or all of these areas. It is imperative, that the effects of AI will be amplified in the coming years, as transportation, finance, manufacturing, retail, healthcare, advocacy, insurance, entertainment, advertising, education and virtually all other sectors of society transform their core processes and models of digital business, taking advantage of intelligent machine learning.

References

1. Silva, F.A.: Smart grid handbook [Book news]. IEEE Ind. Electron. Mag. **12**(1), 59–60 (2018)
2. Kim, T.H., Ramos, C., Mohammed, S.: Smart city and IoT. Future Gener. Comput. Syst., 159–162 (2017)
3. Monteiro, A.C.B., et al.: Development of a laboratory medical algorithm for simultaneous detection and counting of erythrocytes and leukocytes in digital images of a blood smear. In: Deep Learning Techniques for Biomedical and Health Informatics, pp. 165–186. Academic Press (2020)
4. França, R.P., et al.: Potential proposal to improve data transmission in healthcare systems. In: Deep Learning Techniques for Biomedical and Health Informatics, pp. 267–283. Academic Press (2020)
5. Chee, C.Y.K., Tong, L., Steven, G.P.: A review on the modelling of piezoelectric sensors and actuators incorporated in intelligent structures. J. Intell. Mater. Syst. Struct. **9**(1), 3–19 (1998)
6. Al-Turjman, F. (ed.): Artificial Intelligence in IoT. Springer (2019)
7. Yurish, S.Y.: Sensors: smart vs. intelligent. Sens. Transducers **114**(3), I (2010)
8. Jackson, P.C.: Introduction to Artificial Intelligence. Courier Dover Publications (2019)
9. Flasiński, M.: Introduction to Artificial Intelligence. Springer (2016)
10. Akerkar, R.: Introduction to artificial intelligence. In: Artificial Intelligence for Business, pp. 1–18. Springer, Cham (2019)
11. Raj, M., Seamans, R.: Primer on artificial intelligence and robotics. J. Organ. Des. **8**(1), 1–14 (2019)
12. Neto, A.B.L., et al.: A multi-agent system using fuzzy logic applied to eHealth. In: International Symposium on Ambient Intelligence. Springer, Cham (2018)
13. Mittelstadt, B., Russell, C., Wachter, S.: Explaining explanations in AI. In: Proceedings of the Conference on Fairness, Accountability, and Transparency (2019)
14. Amodei, D., et al.: Concrete problems in AI safety. arXiv preprint arXiv:1606.06565 (2016)
15. França, R.P., et al.: Improvement of the transmission of information for ICT techniques through CBEDE methodology. In: Utilizing Educational Data Mining Techniques for Improved Learning: Emerging Research and Opportunities, pp. 13–34. IGI Global (2020)
16. Liu, H., et al.: Foreword to the special issue on recent advances on pattern recognition and artificial intelligence. Neural Comput. Appl. **29**(1), 1–2 (2018)

17. Nadimpalli, M.: Artificial intelligence risks and benefits. Int. J. Inno. Res. Sci. Eng. Technol. **6**(6) (2017)
18. Goodfellow, I., Bengio, Y., Courville, A.: Machine learning basics. In: Deep Learning, vol. 1, pp. 98–164. MIT Press (2016)
19. Simeone, O.: A brief introduction to machine learning for engineers. arXiv preprint arXiv:1709.02840 (2017)
20. Rebala, G., Ravi, A., Churiwala, S.: An Introduction to Machine Learning. Springer (2019)
21. Stamp, M.: Introduction to Machine Learning With Applications in Information Security. CRC Press (2017)
22. Alpaydin, E.: Introduction to Machine Learning. MIT Press (2020)
23. Aksoy, S.: Introduction to Pattern Recognition. Bilkent University (2016)
24. Beyerer, J., Richter, M., Nagel, M.: Pattern Recognition: Introduction, Features, Classifiers and Principles. Walter de Gruyter GmbH & Co KG (2017)
25. Fu, K.-S.: Applications of Pattern Recognition. CRC Press (2019)
26. Voulodimos, A., et al.: Deep learning for computer vision: a brief review. Comput. Intell. Neurosci. **2018**, 1–13 (2018)
27. Odat, E., Shamma, J.S., Claudel, C.: Vehicle classification and speed estimation using combined passive infrared/ultrasonic sensors. IEEE Trans. Intell. Transp. Syst. **19**(5), 1593–1606 (2017)
28. Patterson, J., Gibson, A.: Deep Learning: A Practitioner's Approach. O'Reilly Media, Inc. (2017)
29. Gulli, A., Pal, S.: Deep Learning With Keras. Packt Publishing Ltd (2017)
30. Charniak, E.: Introduction to Deep Learning. The MIT Press (2019)
31. Pattanayak, S.: Introduction to deep-learning concepts and TensorFlow. In: Pro Deep Learning with TensorFlow, pp. 89–152. Apress, Berkeley, CA (2017)
32. Costa-jussà, M.R., et al.: Introduction to the special issue on deep learning approaches for machine translation. Comput. Speech Lang. **46**, 367–373 (2017)
33. Wani, M.A., et al.: Introduction to deep learning. In: Advances in Deep Learning, pp. 1–11. Springer, Singapore (2020)
34. Huang, K., et al. (eds.): Deep Learning: Fundamentals, Theory and Applications, vol. 2. Springer (2019)
35. Li, P., Liu, X.: Common sensors in industrial robots: a review. J. Phys: Conf. Ser. **1267**(1) (2019)
36. Deng, L., Liu, Y.: A joint introduction to natural language processing and to deep learning. In: Deep Learning in Natural Language Processing, pp 1–22. Springer, Singapore, (2018)
37. Deng, L., Liu, Y. (eds.): Deep Learning in Natural Language Processing. Springer (2018)
38. Goyal, P., Pandey, S., Jain, K.: Introduction to natural language processing and deep learning. In: Deep Learning for Natural Language Processing, pp. 1–74. Apress, Berkeley, CA (2018)
39. Goldberg, Y.: Neural network methods for natural language processing. Synth. Lect. Hum. Lang. Technol. **10**(1), 1–309 (2017)
40. França, R.P., et al.: A methodology for improving efficiency in data transmission in healthcare systems. In: Internet of Things for Healthcare Technologies, pp. 49–70. Springer, Singapore
41. França, R.P., et al.: Improved transmission of data and information in intrusion detection environments using the CBEDE methodology. In: Handbook of Research on Intrusion Detection Systems, pp. 26–46. IGI Global (2020)
42. França, R.P., Iano, Y., Monteiro, A.C.B., Arthur, R.: Lower memory consumption for data transmission in smart cloud environments with CBEDE methodology. In: Smart Systems Design, Applications, and Challenges, pp. 216–237. IGI Global (2020)
43. Kaur, P., Sharma, M., Mittal, M.: Big data and machine learning-based secure healthcare framework. Procedia Comput. Sci. **132**, 1049–1059 (2018)
44. Kabalci, E., Kabalci, Y.: Introduction to smart grid architecture. In: Smart Grids and Their Communication Systems, pp. 3–45. Springer, Singapore (2019)
45. Aguilar, M.R., San Román, J.: Introduction to smart polymers and their applications. In: Smart polymers and their applications, pp. 1–11. Woodhead Publishing (2019)

46. Colak, I.: Introduction to smart grid. In: International Smart Grid Workshop and Certificate Program (ISGWCP). IEEE (2016)
47. França, R.P., et al.: An overview of internet of things technology applied on precision agriculture concept. In: Precision Agriculture Technologies for Food Security and Sustainability, pp. 47–70. IGI Global (2021)
48. Crnjac, M., Veža, I., Banduka, N.: From concept to the introduction of industry 4.0. Int. J. Indus. Eng. Manag. **8**, 21 (2017)
49. Zezulka, F., et al.: Industry 4.0–An introduction in the phenomenon. IFAC-PapersOnLine **49**(25), 8–12 (2016)
50. França, R.P., et al.: A proposal based on discrete events for improvement of the transmission channels in cloud environments and Big Data. In: Big Data, IoT, and Machine Learning: Tools and Applications, pp. 185 (2020)
51. Schütze, A., Helwig, N., Schneider, T.: Sensors 4.0–smart sensors and measurement technology enable Industry 4.0. J. Sens. Sens. Syst. **7**(1), 359–371 (2018)
52. Wilkesmann, M., Wilkesmann, U.: Industry 4.0–organizing routines or innovations?. VINE J. Inf. Knowl. Manag. Syst. **48**(2), 238–254 (2018)
53. Lu, Y.: Industry 4.0: a survey on technologies, applications and open research issues. J. Ind. Inf. Integr. **6**, 1–10 (2017)
54. Lee, J., et al.: Industrial artificial intelligence for industry 4.0-based manufacturing systems. Manuf. Lett. **18**, 20–23 (2018)
55. Zheng, P., et al.: Smart manufacturing systems for Industry 4.0: conceptual framework, scenarios, and future perspectives. Front. Mech. Eng. **13**(2), 137–150 (2018)
56. Eifert, T., et al.: Current and future requirements to industrial analytical infrastructure—part 2: smart sensors. Anal. Bioanal. Chem. **412**(9), 2037–2045 (2020)
57. Jarrahi, M.H.: Artificial intelligence and the future of work: human-AI symbiosis in organizational decision making. Bus. Horiz. **61**(4), 577–586 (2018)
58. Semmler, S., Rose, Z.: Artificial intelligence: application today and implications tomorrow. Duke L. Tech. Rev. **16**, 85 (2017)
59. Bundy, A.: Preparing for the future of Artificial Intelligence. AI Soc. **32**(2), 285–287 (2017)
60. Osifeko, M.O., Hancke, G.P., Abu-Mahfouz, A.M.: Artificial intelligence techniques for cognitive sensing in future IoT: state-of-the-art, potentials, and challenges. J. Sens. Actuator Netw. **9**(2), 21 (2020)
61. Silva, Da., Alexandre, R., Silva, F.C.A., Gomes, C.F.S.: O uso do Business Intelligence (BI) em sistema de apoio à tomada de decisão estratégica. Revista GEINTEC-Gestão, Inovação e Tecnologias **6**(1), 2780–2798 (2016)
62. Ravì, D., et al.: Deep learning for health informatics. IEEE J. Biomed. Health Inform. **21**(1), 4–21 (2016)
63. Zhao, R., et al.: Deep learning and its applications to machine health monitoring. Mech. Syst. Sign. Process. **115**, 213–237 (2019)
64. Oh, Y., Park, S., Ye, J.c.: Deep learning COVID-19 features on CXR using limited training data sets. IEEE Trans. Med. Imaging **39**(8), 2688–2700 (2020)
65. Liang, W., et al.: Early triage of critically ill COVID-19 patients using deep learning. Nat. Commun. **11**(1), 1–7 (2020)
66. Khan, S., Yairi, T.: A review on the application of deep learning in system health management. Mech. Syst. Signal Process. **107**, 241–265 (2018)
67. VoPham, T., et al.: Emerging trends in geospatial artificial intelligence (geoAI): potential applications for environmental epidemiology. Environ. Health **17**(1), 40 (2018)
68. Kouziokas, G.N.: An information system for judicial and public administration using artificial intelligence and geospatial data. In: Proceedings of the 21st Pan-Hellenic Conference on Informatics (2017)
69. Wenkel, S.D.: Geospatial Artificial Intelligence. (2019)
70. Yampolskiy, R.V., Spellchecker, M.S.: Artificial intelligence safety and cybersecurity: a timeline of AI failures. arXiv preprint arXiv:1610.07997 (2016)

The Role of Smart Sensors in Smart City

Harpreet Kaur Channi and Raman Kumar

Abstract Smart cities provide critical infrastructure for a network of sensors, cameras, cables, wireless devices, and data centres that allow city authorities to deliver essential services more quickly and efficiently. Intelligent cities also make the use of sustainable construction materials and reduce energy consumption much more environmentally friendly. Practical usage of technology facilitates the construction of an effective transport management program, upgrades healthcare services, and establishes a broad contact network to interact with all businesses, workers, and other governmental interrelationships. The urbanization pattern is rising. Cities around the world face tight budgets and ageing facilities with further population shifts to metropolitan regions. Future communities need to be healthier, resilient, effective, relaxed, engaging, and intelligent. This chapter highlights the need for smart sensors in smart cities for remote control technologies. The smart temperature sensors are elaborated in detail. The applications of smart temperature sensors in smart cities also discussed with examples such as water management system, energy conservation, street lighting system and waste management.

Keywords Smart sensors · Temperature sensors · Light sensors · Smart city · Urbanization

1 Introduction

The urban growth pattern is rising day by day due to the increase in population. So, tomorrow's smart cities need to be cleaner, more secure, effective, convenient, responsive and intelligent [1]. In smart cities, sensor network, wireless devices, and

H. K. Channi (✉)
Department of Electrical Engineering, Chandigarh University, Gharuan, Mohali,
Punjab 140413, India

R. Kumar
Department of Mechanical Engineering, Guru Nanak Dev Engineering College,
Ludhiana, Punjab 141006, India

U. Singh et al. (eds.), *Smart Sensor Networks*, Studies in Big Data 92,
https://doi.org/10.1007/978-3-030-77214-7_2

data centres form a vital infrastructure, allowing communities to provide a critical part of the world's urban growth process. Intelligent buildings can use recycled construction materials to minimize electricity usage. Smart cities are far more eco-friendly. The efficient use of technology helps to establish an adequate transportation services infrastructure, develop health care facilities, and build a broad network of contact between all enterprises, individuals, and beyond regional and sub-national government affairs [2]. The urban infrastructure that interacts with the public can enhance their quality of life in real-time, through traffic safety, garbage management, recycling systems, irrigation systems, parking assistance, alerting local authorities in case of an incident, and enabling the government to keep up-to-date with the city [3]. It is also tempting for both people and industries needs new urban infrastructure. The intelligent network sensors leads to make it more cost-competitive, speedy, and on-scale for prospective communities to incorporate a diverse array of services [4]. Several smart city developments, including Smart Energy Networks, Smart Meters, and ETI, are pilot projects that use smart sensor network.

1.1   Use of Sensor Technology, Remote Control Network, and Smart Mobility

By implementing sensors in the cities, a smart city is willing to build an actual and digital management network for police and local personnel by sharing and making it accessible to the media, the city authorities, companies, and professionals [5]. The sensor framework provides typical data storage system consisting of information from various sensor systems. An automated control system manages both municipalities and suppliers to service providers participating in the project with local data communication infrastructure. The aim is to monitor and think about the regular usage of these networks, chances, and possibilities regardless of public resources [4–6]. The distribution grid can be a service network that involves a storage network, water drainage, a rainwater network, public lighting system, pneumatic waste disposal, climatology, electricity, and indoor home comfort. Informatics and Communication Technology (ICT) changes the way, the societies work, and communication flow facilitates societies longevity, as it provides people with the knowledge, they need to make educated decisions. ICT structures neighbourhoods more effectively and encourages people and local groups to have constructive involvement and an effective support mechanism and promotes engagement with individuals and societies internally and externally. For urban partnerships worldwide, it builds global neighbourhoods [7].

1.2 Smart Sensors

A smart sensor is a device that uses built-in compute resources to conduct prede-fined tasks after identification, and then processes data before transmission. Smart sensors enable more reliable and automatic processing of environmental informa-tion with less noise within the correctly collected data [8]. These instruments are used to track and manage processes across various contexts, including smart net-works, frontline identification, discovery, and a wide variety of research applica-tions. In the internet of things (IoT), the smart sensor is a critical and integral function, that has the ability to relay information through the internet or the same network [9]. They are used to compose a wireless sensor and actuator network (WSAN) with thousands of nodes connected to one or many other sensors, sensor hubs, and individual actuators. Usually, low power compact microprocessors have computational tools. Figure 1 shows a smart sensor made of one form of detector, microprocessor, and communication technologies. The measurement tools must become an integral component of physical architecture. A sensor that sends only data to remote processing is not called an intelligent sensor. These modules may be transducers, amplifiers, analog filters, excitation control, and compensation. The smart sensor also contains software-defined elements that include data transfer, digital transmission, and external system connectivity [10].

1.3 Need and Use of Smart Sensor Technology

In smart cities, IoT technologies use a spectrum of sensors to gather data from a single cloud-based storage resource and transmit it over the internet. The analytics platform that runs on cloud computers reduces the tremendous amount of data provided by user-friendly information and field actuator controls. Sensors are a key component of IoT output, although they are not conventional types that transform physical variables into electrical signals [10, 11]. To play a technically and

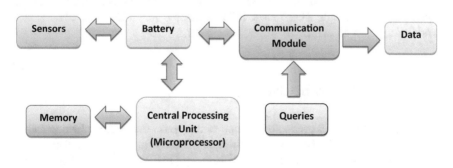

Fig. 1 Architecture of sensor node

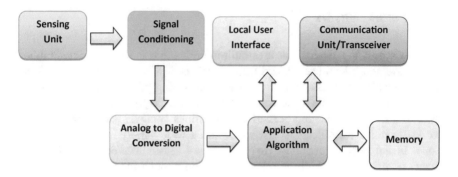

Fig. 2 Smart sensor technology

commercially viable role in the IoT universe, they had to develop into something more creative. As described in the introduction, smart sensors are designed to turn the real-world portion into a digital data stream for transfer into a gateway.

Intelligent sensors are designed as IoT modules, which turn the real-world vector into a digital data stream for gateway transmission, as shown in Fig. 2. An embedded microprocessor unit (MPU) is used to execute the device algorithms. Filtering, compensation and other system-specific signal processing functions may be carried out. However, for many other tasks, the MPU intelligence can be used as a way of reducing the load on the central IoT tools [12]. Calibration data will, for example, be forwarded to the MPU to set the sensor for any changes in output automatically. The MPU can also monitor any output parameters that tend to wander above reasonable expectations. The operators can also take proactive measures in the event of a catastrophe.

The sensor will run in extraordinary reporting mode only if the values of the measured variables differ considerably from previous sample values. This reduces both the core estimation resource load. The smart power demand of the sensor is typically a crucial advantage because, in the absence of linked power, the sensor needs to rely on a battery. If two components are included in the sensor, the smart sensor may also be self-diagnosed. Any emerging drift can be sensed immediately at one of the sensor outputs. Besides, the second measurement factor may be continued, for example, if a sensor fails. A sample may also have two sensors working together to maximize surveillance input.

2 Types of Smart Sensors

The smart sensors yield productive outcomes in diverse markets, so the companies use them significantly. The cost-effectiveness feature of intelligent sensors means that it is being used by healthcare growth industry [13]. The reason for the reduced cost is the prominent use of technology on smart sensors. Reduced repeated

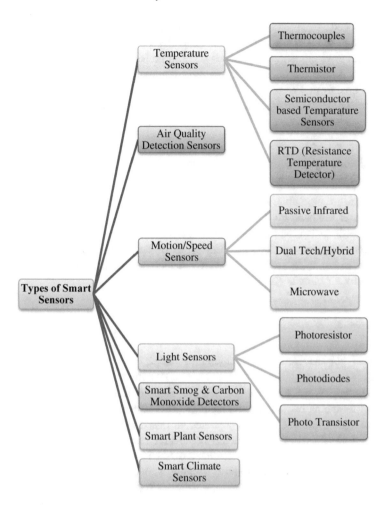

Fig. 3 Types of smart sensors

monitoring is another cause of the cost-effectiveness of smart sensors. Different sensor types are available are shown in Fig. 3.

2.1 Temperature Sensors

An electronic device that measures the temperature and renovates the input information into electronic numbers to record or indicate temperature changes are known as temperature sensors. It works on the principle of the voltage transversely the diode terminals. If the voltage rises, the temperature likewise rises, tracked by a voltage bead amid the base and emitter's transistor terminals in a diode. Non-contact temperature sensors are generally based on infrared sensors. In the up-to-date environment, temperature sensors are one of the most prevalent technologies with the potential to measure the air temperature by collecting various data conversions from Fahrenheit or Celsius classes. There are many smart sensors inside smart thermometers, thermostats, and weather stations [14].

2.2 Types of Temperature Sensors

For building temperature control, water temperature control, and refrigerator control, normal temperature controls are utilized. In many other uses, including the market, medical and industrial electronics, temperature sensors are also critical [15]. In modern electronics, there are four types of temperature sensors most widely used: thermocouples, RTD (resistance detectors), thermistors, and integrated semiconductor circuits (IC).

2.2.1 Thermocouples

The most popular form of a temperature sensor is thermocouple. It is used for automotive, automobile, and consumer applications. Thermal couplings are self-powered, do not need excitement, are capable of working over an extensive range of temperatures, and rapid response times. It triggers the Seebeck influence. This voltage differential can be calculated and used in measuring the height. Many thermocouple forms are constructed of several contents for measuring different temperature range [14, 15]. The Seebeck effect is a physique where two varying conductors create a differential of voltage between the two substances. The numerous styles are characterized by letters. The K form is the most widely encountered. The characteristics of a few popular thermocouples are given in Table 1.

Table 1 Thermocouple types and characteristics

Code type	Conductors alloys (±)	Sensing temperature	Sensitivity (μV/°C)
R/S	Copper/Copper Nickel compensating	−50 to 1750 °C	10
E	Nickel Chromium/Constantan	−40 to 900 °C	68
B	Platinum Rhodium	0 to 1820 °C	10
J	Iron/Constantan	−180 to 800 °C	55
N	Nicrosil/Nisil	−270 to 1300 °C	39
K	Nickel Chromium/Nickel Aluminum	−180 to 1300 °C	41
T	Copper/Constantan	−250 to 400 °C	43

The drawback of thermocouples is that their limited output voltage, requiring precise amplification, sensitivity to outside noise over long wires, and cold intersection, can make temperature measurement difficult. The cold junction is the junction where the copper traces of the signal circuit meet thermocouple wires. Maxim integrated provides digital output thermocouples like MAX31855 and MAX31856, which are also needed to compensate for the so-called cold junction compensation. Signal conditioning can be aided by using a digital converter analogue at high resolution, a low noise quality gain stage, and a cold junction adjustment sensor. These instruments enable thermocouple circuit designers to include specific solutions in the compact bundle for signal conditioning. They are working with various common thermocouple forms.

The smart thermocouple developed based upon recognized sensor recital features to progress industrial temperature measurement and it abridged measurement uncertainty, augmented robustness, and failure forecast and recognition [16]. A high accuracy thermocouple designed and developed as a smart industrial thermometer for data logging and linearization. K-type thermocouple selected with a 12-bit analog-to-digital converter and an 8-bit microcontroller. The developed system can be employed for different types of multidrop sensor network with minor alteration [17]. The moisture monitoring and temperature measuring system are generated based on IoT and thermocouple temperature sensors. The developed system provides all information about the structural health of a building and can evade accidents [18].

2.2.2 RTD (Resistance Temperature Detector)

The resistance of any metal often increases as the temperature changes. The RTD temperature sensors are based upon the difference in resistance. An RTD is a resistor with given characteristics of resistance against temperature. Platinum is often referred to as PRTD, the most common and specific substance used to manufacture RTDs. They are also usable at a resistance of 100 and 1000 kg at 0 °C [19]. It is referred to respectively as PT100 and PT1000. Platinum RTDs are chosen because they give a virtually linear temperature response, reliable and precise responses, and a wide range of temperatures. RTD components typically have a greater thermal mass and react more efficiently to temperature fluctuations than thermal couplings. Their exactness and repeatability mean that the RTDs are mostly used in specific applications. The conditioning of signals at RTDs is critical [20]. They also need a flow of enthusiasm in the RTD. The durability can be measured if current is known. The two-wire choice is helpful where the lead's length is short enough to have little effect on the calculation's precision. A Trois-wire attaches an RTD sensor with the current of energy. This will null the wire resistance. Four-wire is more effective since the influence of wire resistance is reduced by separate force and sound. The MAX31865 provides a dedicated 15-bit RTD signal conditioning circuit with a solution for accelerated designs for PT100 and PT1000 RTDs.

An RTD premeditated and verified through a digital IEEE 1451.2 smart sensor interface standard. The measurement and computational resolution achieved within 2 ppm [21]. Riemer and Davis [22] described that an in-situ temperature sensor for Li-Ion batteries, RTD's have a simple linear response to temperature and can accurately sense from −40 °C through thermal runaway initiation. Li et al. [23] demonstrated the experimental verification of platinum resistance temperature transmitter. It has miniaturization capabilities, decent portability, and high performance.

2.2.3 Thermistors

The thermistor is an amalgamation of thermal and resistor and is a resistance thermometer reliant on temperature. Temperature variations induce observable resistance shifts, which are identical to RTDs. Typically, thermistors are made of a polymer or a ceramic substrate. The thermistors are, in most cases, cheaper, but still less precise than RTDs. In two-wire configurations, most thermistors are available. The thermistor NTC (Negative Timer Coefficient) is the most frequently used thermistor for temperature control. The resistance of an NTC thermistor decreases with the rise of temperature. Thermistors have a relation to non-linear tolerance to temperature. It involves a significant correction to interpret the results accurately. A typical method of using a thermistor consists of separating a thermistor and fixed value resistor, with an ADC digital display [24]. For various practical applications such as non-contact sensation, human skin and wind temperature control, etc., Wu et al. [25] developed a 3D porous Gr-based "self-calibration" thermistor with competitive efficiency. Arathy et al. [26] demonstrated a simple, precise and easy tool for screening and early detection of breast cancer based on the higher temperature of abnormal tissues using excellent reliability chip thermal sensor probes and developed 3D and 2D thermal imaging system that can predict the hot spot precisely by simple surface temperature measurements. The radiation effect on indoor air temperature was studied experimentally by Lundström and Mattsson [27]. They investigated steps using thermocouples and thermistors to reduce the effect on some traditional temperature sensors.

2.2.4 Semiconductor Based Temperature Sensor or Integrated Circuit (IC) Temperature Sensor

It operates with reverse bias and has a trivial capacitance and a little seepage current besides arranged on thin silicon wafers. They are compact, yield in line outputs, and have a minor assortment of temperature. There are two types of semiconductor ICs with temperature sensors available local temperature and remote optic sensor. ICs that can measure their die temperature using the physical properties of the transistor are local temperature sensors.

The temperature sensors measure the transistor's current temperature from a remote digital sensor. Analog outputs could be voltage, current or electrical, while the various formats, such as I2C, SMBUS, 1-Wire, and SPI, could represent digital data. Local temperature sensors notice the temperature on the printed circuit boards or in the atmosphere [28]. A very compact local temperature sensor can be used in various applications and battery-operated applications. The MAX31875 is a software temperature sensor for remote applications. The distinction is that the sensor chip is disconnected from the transistor. A bipolar sensing transistor is used in some microprocessors and FPGAs to test the target IC die-temperature.

The incredible ability of AlGaN/GaN transistors with high electrical conductivity has been demonstrated in the development of solid-state microsensors. Due to its excellent chemical stability, good surface charge sensitivity, high temperature resistance efficiency and low power consumption characteristics, it is used to track pollutants, metal ions, anions, biomolecules and other substances [29]. The metal oxide semiconductor gas sensors were used to measure concentrations of NO_2 and O_3 at an urban background site on a university campus, and these two gases were closely linked to traffic pollution and harmful to human health [30]. Ji et al. [31] demonstrated the ability of semiconductor-based pressure sensors, including remote surgery, health tracking, and robotics, to offer comprehensive applications in next-generation electronics.

2.3   Light Sensors

A light sensor is a photoelectric device that transforms the observed light energy (photons) into electricity. These sensors are furnished with photo-sensitive resistors, which aid to adjust the conductivity of the light it gets. If less light arrives, they will generate high resistance, and more light will provide less resistance. Light sensors, which work on the principle of temperature sensors, are also one of the most used methods these days. In the process of the night-vision mode, however, resistors are connected with a smart security camera to assess voltage, and light sensors serve as reliable burglar controls. In contrast, the camera is visible at night. There are different light sensors available, mainly Photoresistor, Photodiodes, and Phototransistors [32].

2.3.1   Photo-Resistors (LDR)

Photo-resistors, also known as a light-dependent resistor (LDR) the most commonly used in a circuit with a light sensor. The photo-resistors are used to recognize whether lights are on and off, to measure relative light levels over one day. A semiconductor substrate, which is extremely sensitive to visible and near-infrared radiation, is referred to as cadmium sulfide cells. The photo-resistors operate as regular resistors, except the difference in resistance is based on the amount of light to which it is exposed. The high light strength induces decreased cadmium-sulfide cell resistance. The light strength of the cadmium sulfide cells results in higher tolerance. LDR is used in applications such as street lamps, this operating rule can be found in which the higher light strength corresponds to lower resistance, and no light produces [32]. Elwi [33] identified a steep antenna based on a Hilbert-shaped fractal metamaterial with 3×2 arrays printed on a rectangular region and supported with a coaxial waveguide structure suitable for the application of the key of amplitude change. An independent electronic device presented on the basis of the contrast method for sunshine period monitoring and designed to work on a horizontal surface with the aid of four photoresistors arranged at $90°$ in a radius of 20 mm diameter separated by a shading structure used to create a shadow pattern on the detector feature [34].

2.3.2 Photodiodes

Another kind of light sensor is photodiodes. However, it is more difficult to light, quickly turning light into electric current flow instead of utilizing a transition in resistance like LDR. Image tracker, also known as a photodetector. The photodiodes are primarily made of silicon and germanium and have optical filters, built-in lenses, and surface areas. Photodiodes operate according to the functioning theory, the internal photoelectric effect. To put it simply, electrons are loosened by a light beam that leads to electron-holes that allow an electric current to travel through it. The luminous the Sun, the higher the electric current. Since the pictorial current is directly proportionate to the light intensity, it is suitable for light sensing, which involves rapid changes in light response [32, 35]. The clamped photodiode recital analyzed epitaxial wafers with high resistivity for a duration of flight statements. Its usage will greatly improve the performance of crosstalk, enable pixels to function at higher modulation frequencies and allow large pixels to have strong decoding capabilities [36]. A VLP framework has been developed that uses a custom tag that uses several photodiodes. A received signal strength-based fingerprinting is used for indoor localization using a weighted k-nearest neighbor algorithm. It ensures high positioning accuracy and helps the current lighting system to be leveraged [37]. A general approach introduced by ink formulation engineering for printing wavelength-selective bulk-heterojunction photodetectors. This method effectively distinguishes the viscoelastic ink properties from the optical reaction, simplifying process growth [38].

Since infrared light is receptive to photodiodes, it may be used for other purposes. The photodiode functions are as follows:

- Consumer electronics from floppy discs to smoke and even remote controls.
- A photodiode is used in biomedical devices such as calculation and diagnostic equipment/instruments.
- Solar power devices, including solar panels.

2.3.3 Phototransistors

The phototransistor is the kind of light sensor transistors with the base terminal exposed. The photons strike the light and trigger the transistor, instead of transfer current into the base. It is finished with a bipolar semiconductor. A photodiode + amplifier can be defined as the light phototransistor sensor. With the improved enhancement, the phototransistors have far greater light sensitivity. In contrast with photodiodes, it is not significantly improved in low light detection [39]. In phototherapy for skin disorders such as psoriasis, vitiligo, and atopic dermatitis, an ultraviolet-B sensitive thin-film phototransistor has been developed and characterized to track the strength of UV radiation [40]. A new on-chip architecture proposed for low-power consumption phototransistor sensors based on an external stimulus optoelectronic detection mechanism and compliant with CMOS technology [39]. Haider et al. [41] suggested a phototransistor consisting of SLG, semiconductor nanoparticles, and metallic nanowires embedded in an elastic film for use in soft electronic technologies with a rippled gate-tunable ultrahigh responsivity nano stock.

Despite the various kinds of light sensors, this device can also be found in many applications:

- Consumer electronics
- Automobiles
- Agricultural Usages
- Security applications.

2.4 Air Quality Detection Sensors

The air quality sensors are instruments that identify and track air pollutants' appearance in the local area. These sensors are also incredibly useful for indoor and outdoor applications. The air quality is also essential, relative to the temperature or humidity. The air quality sensors can monitor the concentrations of CO_2 and the volatile organic compounds (VOCs) with methane and ammonia-like gases. If the unit's score is lower, than the emissions will arise, and this must be fixed as soon as possible, and the hazard to various pathogens will be eliminated. In general, smart sensors will be able to have a clear image of the home's environment. The most air purifier or cleaner uses the smart air quality sensors combined with thermostats, HVAC systems, and air-conditioners [35]. An IoT-based platform for tracking indoor air quality was developed and demonstrated. It consists of a "Smart-Air" air quality sensor system and a webserver to measure aerosol, VOC, CO, CO_2 and temperature humidity concentrations for monitoring air quality [42]. For constant real-time indoor air quality control, a prototype called the ICA system has been launched. In order to track the concentration of radon, CO_2, CO, VOCs and meteorological parameters, such as temperature, pressure and relative humidity, the

ICA system presents sensors [43]. Insights provided into new important indicators for air pollution measurements such as PM 2.5 and VOC sensing technologies that enable personal air quality monitoring to improve people's health and well-being [44].

2.5   Motion/Speed Sensors

The use of passive infrared sensor and BlueTooth device were discussed to notify a house occupant of a guest coming into the house to protect their residence from unauthorized people and as such, setting up an effective and reliable presence detection for residential and office usage [45]. By detecting the difference in infrared radiation around the polarity of the sensor, the passive infrared (PIR) sensors detect human activity to artificially trigger the motion required for static human detection [46]. The monitoring of bus passengers with the help of IoT system was described which uses PIR sensors with mobile apps that can only detect movements made by humans alone to overcome human counting errors and the accuracy of objects detected by sensors [47].

2.5.1   Dual Tech/Hybrid

For difficult real-world smartphone verification, a hybrid deep learning method was suggested. It guarantees accessibility, stability and the safety of private data on mobile devices without disrupting users to resolve the set of problem solutions that prevail. The problems include poor de-noising capability, inadequate readiness, and low coverage of feature extraction [48].

2.5.2   Microwave Motion Sensors

Paredes et al. [49] have developed, produced and validated a prototype for the encoder/reader of an electromagnetic encoder-based microwave device useful for gesture recognition applications.

The adaptable assertions of microwave sensors in large and small-scale human movements were shown by Liu and Chen [50]. Finger stretching, heart pounding and vocal sounds etc. are included. In areas of human–machine interaction, health care and connected technology, the advanced versatile and adaptive sensing content has demonstrated tremendous promise. The piezoresistive behavior of the micro-wave-cured sensors were demonstrated with the bottommost CNT concentration by noticing human motions such as sitting and standing, swaying, and grabbing an object [51].

2.5.3 Motion Sensors

Compared with the recent smart motion sensors, most of the previous motion sensors were not so useful. It was not suitable for monitoring, but the new iteration is expected to yield exact results. Technological advancements, customizable size, and sensitivity settings have improved significantly by applying the smart movement sensors. In the smart movement sensors, the bulk of the activities are performed by measuring a protected environment with infrared rays and activated during detection. Motion sensors are now mounted in safety cameras to increase security [52]. The use of passive infrared sensor and BlueTooth device were discussed to notify a house occupant of a guest coming into the house to protect their residence from unauthorized people and as such, setting up an effective and reliable presence detection for residential and office usage [45]. By detecting the difference in infrared radiation around the polarity of the sensor, the passive infrared (PIR) sensors detect human presence to artificially trigger the motion needed for steady human detection [46]. The monitoring of bus passengers with the help of IoT system was described which uses PIR sensors with mobile apps that can only detect movements made by humans alone to overcome human counting errors and the accuracy of objects detected by sensors [47].

2.6 Smart Smog and Carbon Monoxide Detectors

The amount of pollutant contaminants in the air has risen significantly, and pollution has become typical for developing countries. Such chemicals indeed cause many illnesses, and children are affected by many diseases, as urban affluence is heavily polluting. There are two types of smog and carbon monoxide detectors, where the light sensor is mounted in a photoelectric light source at a 90-degree angle. Essentially, the light does not come in contact with the sensor and extends the light beam through the smoke passage, which activates the sensors to produce accuracy performance. The smoke is responsible for breaking the circuit. These responsive smog detectors and carbon monoxide detectors may also alert the local fire alarms to take the samples they need [52]. Kodali et al. [53] explained that MQ135 air quality sensor could be used to monitor the pollution of indoors. It sends alerts to the user by using the Internet when the measured concentration of pollutants goes beyond safer level.

2.7 Smart Plant Sensors

Plants are an indispensable portion of vigorous life and need to develop the planting portion of our region and maintain a healthy climate. By cooperating with intelligent plant sensors, people can do what they need to survive on plants such as the required light, temperature, land, and soil for the development and fertility processes. Humidity is also an essential factor in the broad plantation and can be measured by electricity transmission from the soil [54]. The above elements would also be a considerable benefit during field cultivation and enhance the household's health. People can send the data directly through the smart plant sensors activated with different mobile apps to our smartphone. A smart plant-wearable biosensor has been developed by Zhao et al., which can be used for in-situ analysis of organophosphorus pesticides on crop surfaces to satisfy the need for rapid and non-destructive identification for the potential growth of precision agriculture [55]. Rus et al. [56] proposed a smart plant management system to monitor the irrigation and fertilization of plants based on their optimum growing conditions, varying from temperature to soil moisture, as additional fertilization and water requirements are required for various types of plants.

2.8   Smart Climate Sensors

In the intelligent era, people need to become smart and technical advancement can't go backward. Multiple mobile devices are introduced in a contemporary scenario that is internet-linked and can relay the data of the item's relevant activities to the cell phone 24/7. It helps smart climate sensors to adjust timetable accordingly, so those innovative systems often make human wiser. Since they have been paired with a smartphone, warnings can be forwarded through the daily temperature [57].

3   The Usage of Smart Sensors: Prominent Examples

Smart sensors improve the ability to track and report on the world around people. They work in almost all industries to make human existence more straightforward and much healthier. Smart sensors are changing lighting to the tune of moods, turning on devices such as water heaters, maintaining protection, monitoring devices, and much more. Smart sensors allow greater transparency of business workflows, recognize everyday working habits and evaluate the facilities' environmental conditions. Smart sensors also enable enterprise management to track, automate and increase operating performance. With the IoT applications, which penetrate more of life than ever before, precise sensors are rising significantly. The IoT cannot function without smart sensors and is a critical component of IoT networks and smart cities. The smart sensor usage examples are discussed here.

3.1 Smart Water Management (SWM) Using Smart Sensors

For its healthy use, water is an essential natural resource that demands continuous quality control. The detection of water pollution has historically been conducted manually, gathering and testing water samples in laboratories since these approaches do not supply real-time data. A water quality monitoring system based on the wireless sensor network to track water quality data is a good example. The system architecture relies on the topology of the hierarchy in which the situation is split into four general regions. Each of these classes consists of multiple wireless sensor nodes responsible for sensing, processing, and connectivity. Three primary parameter controls influence water quality, i.e., pH, conductivity, and water temperature built for the system's wireless sensor node. The sensor node architecture consists primarily of a module for signal conditioning; a processing module implemented using a PIC microcontroller, and a Zigbee radio wireless communication module [58, 59]. Sensed parameter values will then be transmitted to the base station in real-time using Zigbee communication after the required signal conditioning and processing techniques. This device provides an energy-efficient and economical sensor unit using affordable low-power hardware construction equipment to track water consistency [60].

To minimize operating and energy costs, speeding up the water cycle, and to reduce water pollution in the environment, SWM is concerned mainly with the optimization of water distributions and sewers. ICT dependent sensors and communications technologies must provide the required monitoring and control (real-time) of the system of water delivery. It includes pressure and flow meters, power usage, water temperature, water use, leakage detection, sensors, and meters. The communication infrastructure, such as network sensor cellular, GPS/GPRS, wireless LAN, meter technology. Monitor and control systems carry out various functions from data acquisition by sensors, network connection across remote sites, central computer data merge (master unit), and operator terminals/workstations are monitoring [58, 59, 61].

3.2 Energy Conservation Using Smart Sensors

It is vital that how smart sensors and data collection can effectively handle energy. Rising energy prices are a noteworthy problem everywhere for households. Surplus power consumption is not positive either to the surroundings or to the power bank. Sensors have now activated the "Advised Metering Infrastructure (AMI)" idea supporting cities' energy control. Cities suggest using the digital meters installed with the phase measurement unit sensors and communications module, allowing bidirectional interactions between customer and supplier [62]. It helps to verify the meter status in response to a customer call before sending a repair team. This measure prohibits the shipping of unwanted field personnel to customer sites. It can provide users with energy consumption information in real-time so that a person can very readily comprehend. Based on these results, consumers can change priorities and make informed choices about their use without waiting for the end of a month for their energy bill. Without any manual means, the energy usage must be regulated by machines, considering how much people fail to turn on and off the electrical devices. For example, imagine sitting back at home and working on smart power ideas in our name. Many people go to bed, and the computer remains on the whole night while viewing a TV [63]. The motion of a person who goes to sleep, which automatically shuts the TV off after a while, can be sensed with intelligent sensor controls. A successful energy conservation strategy includes the use of computer programmers and sensors for TV viewing.

Thermostats wirelessly shared over the internet allows people to handle cooling and warming alerts with a fingertip. With a broad range of sensors available to detect different parameters such as human activity, temperature, pressure, and activity of certain gases and much more, it is also essential to use this IoT driving technology to improve construction automation. The new smart thermostats can be efficiently operated using mobile gadgets such as tablets and smartphones. The value of the sensors in the carbon dioxide transmitter mix is a good case of use. The CO_2 relay transmitter produces heat automatically. And each stretch a person goes into and out of the apartment, it will feel and turn on and off automatically. This simple HVAR method will save a lot of resources. When human beings enter a room, sensors can sense temperature changes and schedule intelligent lights to be triggered. On the internet of advanced technologies, interconnected sensors and autonomous devices can create several gadgets and functions.

IoT is now being used to enhance plant productivity, the efficiency of indoor air, and increase the flexibility of cooling systems, trash compactors, and solar rooftops. The emerging smart sensors, including wireless sensors, nano-sensors, MEMS sensors, etc., demonstrate the way towards a vital future. As for the energy supplies, IoT has also implemented new methods of providing, tracking, evaluating, and pricing electricity based on many parameters. Smart meter automation for data processing concerning regional consumption patterns and developments and other demographics has contributed to an exponential decrease in the time and human resources needed to obtain data from the meters. This data from sensors can now be conveniently sent to the cloud for unified analysis and visualization that enables energy suppliers to guarantee demand control and continuous supply and adjust their price structure according to consumer trends and peak loads. For energy users and suppliers, this is a perfect win–win IoT rollout situation [64, 65].

3.2.1 Benefits of IoT and Smart Sensors for Energy Efficiency

- Sensor energy-saving activities.
- The impact of global warming has decreased

- No manual operation required.
- Device for sensing human activity.
- Multifunctional wireless device monitoring.
- Regulation of internal light decrease.
- Enhanced functions of existing electrical and cable networks.

With the implementation of a sensor-integrated smart meter as intelligent bottoms, power distributors gain when adequately used in combination with cloud enablement and analysis. The total energy consumption and energy consumption paradigm have now changed with predictive modeling demand and production response, costs, and integrated resource management. Simply put, our ultimate goal is to reduce energy consumption without disrupting our everyday lives. An immense volume of data from sensors is gathered, which is versatile enough to guide interventions that boost efficiency and cost-effectiveness [66]. The developing environment needs intelligent, self-adaptive solutions, which follow the needs and provide maximum advantages.

3.3 Smart Sensor Street Lighting Control System

Smart street lighting refers to street lighting, which adapts for pedestrian, bike and automobile movement. Smart street lighting, also known as intelligent street lighting, dims if no motion is observed but illuminates when movement is detected [66]. This lighting form varies from conventional, fixed or dimmable street lighting, which diminishes at preset times. An intelligent streetlight regulation motion sensor that unlocks automatically when it detects a vehicle or a pedestrian in the field. If no operation is carried out in the field, the light is automatically adapted to a minimum light level controlled. This system has many benefits, such as substantial energy savings, where light is required, enhanced LED lamp life, lower light, and more profound sky emissions [67].

3.4 Smart Waste Management with Smart Sensors

Waste or disposal is the full set of practices and acts needed for waste treatment from start to finish disposal. It involves processing, transportation, treatment, and waste management, along with surveillance and control [67]. It also concerns the legislative and administrative structure for waste management and provides recycling guidelines. Waste systems can cause environmental issues. Although it seems paradoxical, few devices perform better than waste management. The inefficient systems of frequent loads, large trucks, and fitting waste collectors lead to high prices, an enormous environmental effect, and lots of technical and human downtime [68].

Technology-based on smart sensors is a practical solution to waste management. Smart sensor IoT provides waste treatment firms and their users with multiple opportunities. IoT level sensors simplify and refine waste collection processes to save resources and make it greener for companies. To optimize their stock of litter and recycles, a rising number of organizations are using intelligent bins and innovative technology. At present, these technologies use fill stages to display bin completion over time. These data are presented as real-time data to allow bin-service systems and transport routes to optimize performance and quality [69, 70].

4 Concluding Remarks

The chapter describes the role of smart sensing in smart cities. Various smart sensors like air quality detection sensors, temperature sensors, motion/speed sensors, smart smog and carbon monoxide detectors, smart plant sensors and smart climate sensors were discussed along with their working principle and benefits. A variety of applications have been identified in intelligent cities with sophisticated, intelligent sensing capabilities. This include water, recycling of electricity, street lighting and waste management.

The smart city's latest innovations include smart sensor technology due to its ability, importance, and wide range of applications. These new devices are a conceivable novel wave of self-awareness and identification capabilities, crucial to future smart city devices. The smart sensors are operating with increasingly varied specific inputs in micro-electro-mechanical systems. Smart sensors' core functions are dynamic and multi-stage processes, such as raw data acquisition, sensitivity and filtration adaptation, motion sensing, interpretation, and communications. Smart sensors have a full role in every field of smart city life, ranging from HVAC to traffic control, air conditioning systems, and farming. The role of smart sensors in food safety and biological hazard identification, fire risk management and alert systems, environmental surveillance, health control, medical evaluation, industrial and aerospace applications, smart aerial antennas, vehicles, and innovative highways have a significant impact in smart city implementation projects.

Sensing is at the core of smart structures, which can screen itself and perceptually transform on its own. The use of sensors for tracking public facilities in intelligent communities, including bridges, highways and houses, allows the more competent use of capital on the basis of knowledge gathered by sensors. This chapter expressively demonstrated smart sensors working principles that can be implemented in real-time monitoring to eliminate the need for regularly scheduled inspections. The smart sensors can reduce costs and energy consumption in a smart city that also allows for accurate load forecasting. The traffic tracking sensors used on roads gather data required to develop smart transport systems. While current structures and programs are now in place for intelligent communities, infrastructure to incorporate an all-embracing city would remain expensive. Furthermore, there would be a substantially higher cost of restructuring existing towns to suit the universal city model. In comparison, intelligent sensors in the Public Sector are important because they can enhance the method of distribution. Municipalities and communities should take advantage of smart technology and provide infrastructure optimization, such as parking, cultural control, illumination, maintenance, and surveillance. Future scientific experiments may use a hybrid approach of researching intelligent sensory variables and integrating qualitative and quantitative approaches.

5 Future Scope

Potential outcomes of New Sensor Technology Affecting the Future of Smart Cities, estimates that digitalization and advances in the Internet of Things (IoT) are accelerating the large-scale deployment of sensor technologies through cities. Combined with core enabling technology such as artificial intelligence (AI) and high-speed Internet networks, interconnected urban sensor networks are pushing the development of a connected city environment in order to allow efficient use of public capital. Future scope of study covers acoustic, lidar, radar, 3D camera sensors, weather sensors, flow sensors, gas sensors, and temperature and humidity sensors, as well as adaptation scenarios in core smart cities around the globe. Applications such as food security and bio hazard detection, safety prediction and warning, environmental tracking, health and medical Diagnostics and fabrication and aerospace products, smart antennas, cars and smart roads will be deeply affecting smart sensor systems in future.

References

1. Habibzadeh, H., Qin, Z., Soyata, T., Kantarci, B.: Large-scale distributed dedicated- and non-dedicated smart city sensing systems. IEEE Sens. J. 17(23), 7649–7658 (2017)
2. Anjomshoaa, A., Duarte, F., Rennings, D., Matarazzo, T.J., deSouza, P., Ratti, C.: City scanner: building and scheduling a mobile sensing platform for smart city services. IEEE Internet Things J. 5(6), 4567–4579 (2018)
3. Wong, M.S., Wang, T., Ho, H.C., Kwok, C.Y.T., Lu, K., Abbas, S.: Towards a smart city: development and application of an improved integrated environmental monitoring system. Sustainability 10(3), 623 (2018)
4. Soyata, T., Habibzadeh, H., Ekenna, C., Nussbaum, B., Lozano, J.: Smart city in crisis: technology and policy concerns. Sustain. Urban Areas 50, 101566 (2019)
5. Shelton, T., Zook, M., Wiig, A.: The 'actually existing smart city.' Camb. J. Reg. Econ. Soc. 8(1), 13–25 (2014)
6. D'Ignazio, C., Gordon, E., Christoforetti, E.: Sensors and civics: toward a community-centered smart city. In: Paolo, C., Cesare Di, F., Rob, K. (eds) The Right to the Smart City, pp. 113–124. Emerald Publishing Limited (2019)
7. Ismagilova, E., Hughes, L., Rana, N.P., Dwivedi, Y.K.: Security, privacy and risks within smart cities: literature review and development of a smart city interaction framework. Inf. Syst. Front. (2020)

8. Papadavid, G., Hadjimitsis, D., Fedra, K., Michaelides, S.: Smart management and irrigation demand monitoring in Cyprus, using remote sensing and water resources simulation and optimization. Adv. Geosci. **30**, 31–37 (2011)
9. Zhang, Y., Gu, Y., Vlatkovic, V., Wang, X.: Progress of smart sensor and smart sensor networks. In: Fifth World Congress on Intelligent Control and Automation (IEEE Cat. No. 04EX788) IEEE. June 2004 vol. 4, pp. 3600–3606 (2004)
10. Singaravelan, A., Kowsalya, M.: Design and implementation of standby power saving smart socket with wireless sensor network. Procedia Comput. Sci. **92**, 305–310 (2016)
11. Fernandez-Montes, A., Gonzalez-Abril, L., Ortega, J.A., Morente, F.V.: A study on saving energy in artificial lighting by making smart use of wireless sensor networks and actuators. IEEE Netw. **23**(6), 16–20 (2009)
12. 14 E.: Smart Sensors—Overview and Latest Technology (2020). Accessed 15 October 2020.
13. Gervais-Ducouret, S.: Next smart sensors generation. In: 2011 IEEE Sensors Applications Symposium, 22–24 Feb 2011, pp 193–196 (2011)
14. Kunzelman, J., Chung, T., Mather, P.T., Weder, C.: Shape memory polymers with built-in threshold temperature sensors. J. Mater. Chem. **18**(10), 1082–1086
15. Göpel, W., Reinhardt, G., Rösch, M.: Trends in the development of solid state amperometric and potentiometric high temperature sensors. Solid State Ionics **136–137**, 519–531 (2000)
16. Schuh, B.: Smart thermocouple system for industrial temperature measurement. In: SIcon/01. Sensors for Industry Conference. Proceedings of the First ISA/IEEE. Sensors for Industry Conference (Cat. No. 01EX459), 7–7 Nov 2001, pp. 8–11 (2001)
17. Sarma, U., Boruah, P.K.: Design and development of a high precision thermocouple based smart industrial thermometer with on line linearisation and data logging feature. Measurement **43**(10), 1589–1594 (2010)
18. Gupta, R.K.: IoT based a smart sensor for concrete temperature and humidity measurement. Wutan Huatan Jisuan Jishu **26**(10), 42–49 (2019)
19. Blasdel, N.J., Wujcik, E.K., Carletta, J.E., Lee, K., Monty, C.N.: Fabric nanocomposite resistance temperature detector. IEEE Sens. J. **15**(1), 300–306 (2015)
20. Zhang, B., Kahrizi, M.: High-Temperature Resistance Fiber Bragg Grating Temperature Sensor Fabrication. IEEE Sens. J. **7**(4), 586–591 (2007)
21. Wobschall, D., Poh, W.S.: A smart RTD temperature sensor with a prototype IEEE 1451.2 internet interface. In: Proceedings of the ISA/IEEE Sensors for Industry Conference, 27–29 Jan 2004, pp. 183–186 (2004)
22. Riemer, D.P., Davis, M.W.: Evaluation of a manufacturable in-situ thin film RTD temperature sensor for lithium-ion batteries. ECS Meet. Abs. **MA2020–01**(2), 434 (2020)
23. Li, X., Dong, H., Chen, H.: Design of a measurement error calibration device for platinum resistance temperature transmitter. In: 2020 IEEE International Conference on Advances in Electrical Engineering and Computer Applications (AEECA), 25–27 Aug 2020, pp. 602–606 (2020)
24. Gums, J.: Types of temperature sensors. Digi-Key Electronics (2018). Accessed 20 June 2020
25. Wu, J., Yang, X., Ding, H., Wei, Y., Wu, Z., Tao, K., Yang, B.-R., Liu, C., Wang, X., Feng, S., Xie, X.: Ultrahigh sensitivity of flexible thermistors based on 3D porous graphene characterized by imbedded microheaters. Adv. Electron. Mater. **6**(8), 2000451 (2020)
26. Arathy, K., Ansari, S., Malini, K.: High reliability thermistor probes for early detection of breast cancer using skin contact thermometry with thermal imaging. Mater. Express **10**(5), 620–628 (2020)
27. Lundström, H., Mattsson, M.: Radiation influence on indoor air temperature sensors: experimental evaluation of measurement errors and improvement methods. Exp. Thermal Fluid Sci. **115**, 110082 (2020)
28. Voorthuyzen, J.A., Bergveld, P., Sprenkels, A.J.: Semiconductor-based electret sensors for sound and pressure. IEEE Trans. Electr. Insul. **24**(2), 267–276 (1989)
29. Guo, H., Jia, X., Dong, Y., Ye, J., Chen, D., Zhang, R., Zheng, Y.: Applications of AlGaN/GaN high electron mobility transistor-based sensors in water quality monitoring. Semicond. Sci. Technol. **35**(12), 123001 (2020)

30. Peterson, P.J.: Theory and practice of the use of metal oxide semiconductor pollution sensors. University of Leicester (2020)
31. Ji, S., Jang, J., Hwang, J.C., Lee, Y., Lee, J.-H., Park, J.-U.: Amorphous oxide semiconductor transistors with air dielectrics for transparent and wearable pressure sensor arrays. Adv. Mater. Technol. **5**(2), 1900928 (2020)
32. Pitigoi-Aron, R., Forke, U., Viala, R.: Diode-based light sensors and methods. Google Patents (2007)
33. Elwi, T.A.: Remotely controlled reconfigurable antenna for modern 5G networks applications. Microw. Opt. Technol. Lett. (2020)
34. Rocha, Á.B., Fernandes, E.M., Dos Santos, C.A., Diniz, J.M., Junior, W.F.: Development and validation of an autonomous system for measurement of sunshine duration. Sensors **20**(16), 4606 (2020)
35. Conrad, K.S., Manahan, C.C., Crane, B.R.: Photochemistry of flavoprotein light sensors. Nat. Chem. Biol. **10**(10), 801–809 (2014)
36. Fang, X., Xu, Y., Yang, J., Wu, K.: Analyses of pinned photodiodes with high resistivity epitaxial layer for indirect time-of-flight applications. IEEE Access **8**, 187575–187583 (2020)
37. Bakar, A.H.A., Glass, T., Tee, H.Y., Alam, F., Legg, M.: Accurate visible light positioning using multiple photodiode receiver and machine learning. IEEE Trans. Instrum. Meas. **70**, 1–12 (2020)
38. Strobel, N., Droseros, N., Köntges, W., Seiberlich, M., Pietsch, M., Schlisske, S., Lindheimer, F., Schröder, R.R., Lemmer, U., Pfannmöller, M., Banerji, N., Hernandez-Sosa, G.: Color-selective printed organic photodiodes for filterless multichannel visible light communication. Adv. Mater. **32**(12), 1908258 (2020)
39. Li, G., Ma, Z., You, C., Huang, G., Song, E., Pan, R., Zhu, H., Xin, J., Xu, B., Lee, T., An, Z., Di, Z., Mei, Y.: Silicon nanomembrane phototransistor flipped with multifunctional sensors toward smart digital dust. Sci. Adv. **6**(18), eaaz6511 (2020)
40. Singh, A.K., Chourasia, N.K., Pal, B.N., Pandey, A., Chakrabarti, P.: Low operating voltage solution processed (Li$_2$ZnO$_2$) dielectric and (SnO$_2$) channel-based medium wave UV-B phototransistor for application in phototherapy. IEEE Trans. Electron. Dev. **67**(5), 2028–2034 (2020)
41. Haider, G., Wang, Y.-H., Sonia, F.J., Chiang, C.-W., Frank, O., Vejpravova, J., Kalbáč, M., Chen, Y.-F.: Rippled metallic-nanowire/graphene/semiconductor nanostack for a gate-tunable ultrahigh-performance stretchable phototransistor. Adv. Optical Mater. **8**(19), 2000859 (2020)
42. Jo, J., Jo, B., Kim, J., Kim, S., Han, W.: Development of an IoT-based indoor air quality monitoring platform. J. Sens. **2020**, 8749764 (2020)
43. Tunyagi, A., Dicu, T., Cucos, A., Burghele, B., Dobrei, G., Lupulescu, A., Moldovan, M., Niță, D., Papp, B., Pap, I.: An innovative system for monitoring radon and indoor air quality. Rom. J. Phys. **65**, 803 (2020)
44. Herrmann, A., Fix, R.: Air quality measurement based on advanced PM2. 5 and VOC sensor technologies. Sensors Transducers **243**(4), 1–5 (2020)
45. Adewusi, M., Samuel, T., Ayoade, E., Adewale, M.: Passive infrared motion detection with bluetooth interface. Engineers Forum.com.ng (2020)
46. Andrews, J., Kowsika, M., Vakil, A., Li, J.: A motion induced passive infrared (PIR) sensor for stationary human occupancy detection. In: 2020 IEEE/ION Position, Location and Navigation Symposium (PLANS), pp. 1295–1304. IEEE (2020)
47. Rahmatulloh, A., Nursuwars, F.M.S., Darmawan, I., Febrizki, G.: Applied Internet of Things (IoT): the prototype bus passenger monitoring system using PIR sensor. In: 2020 8th International Conference on Information and Communication Technology (ICoICT), 24–26 June 2020, pp. 1–6 (2020)
48. Zhu, T., Weng, Z., Chen, G., Fu, L.: A hybrid deep learning system for real-world mobile user authentication using motion sensors. Sensors **20**(14), 3876 (2020)
49. Paredes, F., Herrojo, C., Martín, F.: Microwave encoders with synchronous reading and direction detection for motion control applications. In: 2020 IEEE/MTT-S International Microwave Symposium (IMS), 4–6 Aug 2020, pp. 472–475 (2020)

50. Liu, P., Chen, W.: Microwave-assisted selective heating to rapidly construct a nano-cracked hollow sponge for stretch sensing. J. Mater. Chem. C **8**(27), 9391–9400 (2020)
51. Herren, B., Saha, M.C., Liu, Y.: Carbon nanotube-based piezoresistive sensors fabricated by microwave irradiation. Adv. Eng. Mater. **22**(2), 1901068 (2020)
52. Faglia, G., Comini, E., Pardo, M., Taroni, A., Cardinali, G., Nicoletti, S., Sberveglieri, G.: Micromachined gas sensors for environmental pollutants. Microsyst. Technol. **6**(2), 54–59 (1999)
53. Kodali, R.K., Pathuri, S., Rajnarayanan, S.C.: Informatics (2020) Smart indoor air pollution monitoring station. In: International Conference on Computer Communication and Informatics (ICCCI), pp. 1–5
54. Giraldo, J.P., Wu, H., Newkirk, G.M., Kruss, S.: Nanobiotechnology approaches for engineering smart plant sensors. Nat. Nanotechnol. **14**(6), 541–553 (2019)
55. Zhao, F., He, J., Li, X., Bai, Y., Ying, Y., Ping, J.: Smart plant-wearable biosensor for in-situ pesticide analysis. Biosens. Bioelectron. **170**, 112636 (2020)
56. Rus, A.C., Khan, M.R.B., Ali, A.M.M., Billah, M.M.: Optimal plant management system via automated watering and fertilization. AIP Conf. Proc. **2233**(1), 050013 (2020)
57. Antonacci, A., Arduini, F., Moscone, D., Palleschi, G., Scognamiglio, V.: Nanostructured (Bio)sensors for smart agriculture. TrAC, Trends Anal. Chem. **98**, 95–103 (2018)
58. Fangmeier, D., Garrot, D., Mancino, C., Husman, S.H.J.A.P.: Automated irrigation systems using plant and soil sensors. ASAE Publication **4–90**, 533–537 (1990)
59. Domoney, W.F., Ramli, N., Alarefi, S., Walker, S.D.: Smart city solutions to water management using self-powered, low-cost, water sensors and apache spark data aggregation. In: 2015 3rd International Renewable and Sustainable Energy Conference (IRSEC), 10–13 Dec 2015, pp. 1–4 (2015)
60. Park, J., Kim, K.T., Lee, W.H.J.W.: Recent advances in information and communications technology (ICT) and sensor technology for monitoring water quality. Water **12**(2), 510 (2020)
61. Mohammed Shahanas, K., Bagavathi Sivakumar, P.: Framework for a smart water management system in the context of smart city initiatives in india. Procedia Comput. Sci. **92**, 142–147 (2016)
62. Wang, X., Ma, J.-J., Wang, S., Bi, D.-W.J.S.: Prediction-based dynamic energy management in wireless sensor networks. Sensors **7**(3), 251–266 (2007)
63. Siva Ranjani, S., Radha Krishnan, S., Thangaraj, C., Vimala Devi, K.: Achieving energy conservation by cluster based data aggregation in wireless sensor networks. Wireless Pers. Commun. **73**(3), 731–751 (2013)
64. Jazizadeh, F., Kavulya, G., Kwak, J.-Y., Becerik-Gerber, B., Tambe, M., Wood, W.: Human-building interaction for energy conservation in office buildings. In: Construction Research Congress 2012, pp. 1830–1839 (2012)
65. Liu, R.P., Rogers, G., Zhou, S.: WSN14–3: Honeycomb architecture for energy conservation in wireless sensor networks. In: IEEE Globecom 2006, 27 Nov–1 Dec 2006, pp. 1–5 (2006)
66. Guiling, W., Irwin, M.J., Berman, P., Haoying, F., Porta, T.L.: Optimizing sensor movement planning for energy efficiency. In: Proceedings of the 2005 International Symposium on Low Power Electronics and Design ISLPED '05, 8–10 Aug 2005, pp. 215–220 (2005)
67. Nanavati, K., Prajapati, H.K., Pandav, H., Umaria, K., Desai, N.K.: Smart autonomous street light control system (2016)
68. Longhi, S., Marzioni, D., Alidori, E., Buo, G.D., Prist, M., Grisostomi, M., Pirro, M.: Solid waste management architecture using wireless sensor network technology. In: 2012 5th International Conference on New Technologies, Mobility and Security (NTMS), 7–10 May 2012, pp. 1–5 (2012)
69. Wijaya, A.S., Zainuddin, Z., Niswar, M.: Design a smart waste bin for smart waste management. In: 2017 5th International Conference on Instrumentation, Control, and Automation (ICA), 9–11 Aug 2017, pp. 62–66 (2017)
70. Folianto, F., Low, Y.S., Yeow, W.L.: Smartbin: smart waste management system. In: 2015 IEEE Tenth International Conference on Intelligent Sensors, Sensor Networks and Information Processing (ISSNIP), 7–9 April 2015, pp. 1–2 (2015)

Impact of AI and Machine Learning in Smart Sensor Networks

Impact of AI and Machine Learning in Smart Sensor Networks for Health Care

S. Kaja Mohideen, Latha Tamilselvan, Kavitha Subramaniam, and G. Kavitha

Abstract Recently, one of the evolving technologies that has huge influence in the arena of research is Wireless Smart Sensor Networks (WSSN). The WSSN is fortified with an arrangement which incorporates components for observing, calculating and communication. These networks can perceive and respond to procedures or occurrences in an indicated ambience. WSSN signifies the subsequent evolutionary expansion stage in engineering like ecological observation, industrial mechanization, traffic observation and robot control. WSSN has several exclusive features like power ingestion, dimension, low cost, scalability, agility and elasticity. Artificial Intelligence (AI) is the replication of human acumen procedures by computer systems. Of late, AI procedures have been used efficaciously to resolve issues in engineering. AI comprises of skilled systems, speech identification, machine learning, deep learning platforms and robotic process automation. Machine Learning (ML) has been propelled as an exclusive technique for AI. ML can be demarcated as the learning procedures for enhancement of computer models that can augment the act of systems. Recently, use of ML technologies in automation of health care systems has been proficient. For diagnosing the diseases and predicting the risks of critical diseases in advance, ML could be applied in Internet of Things (IoT) based WSN. The ML comprises of supervised, semi supervised, unsupervised

S. K. Mohideen (✉)
Department of Electronics and Communication, School of Electrical and Communication Sciences, B. S. Abdur Rahman Crescent Institute of Science and Technology, Chennai, Tamilnadu, India
e-mail: kajamohideen@crescent.education

L. Tamilselvan · K. Subramaniam · G. Kavitha
Department of Information Technology, School of Computer, Information and Mathematical Sciences, B. S. Abdur Rahman Crescent Institute of Science and Technology, Chennai, Tamilnadu, India
e-mail: latha.tamil@crescent.education

K. Subramaniam
e-mail: skavitha.jegan@gmail.com

G. Kavitha
e-mail: gkavitha.78@crescent.education

and reinforcement learning methods. This chapter offers an exclusive study of machine learning approaches utilised on WSSN which can be applied in health care systems.

Keywords Smart sensor network · Artificial intelligence · Machine learning · Health care

1 Introduction

Wireless Sensor Network (WSN) is a network of sensors dispersed spatially and used for monitoring the environmental conditions and transfer the collected data to a predetermined location for further processing. In WSN the sensors have no direct connection to the internet instead they interact with the external world through router or central node. One of the sprouting technologies that has enormous impact in the field of data communication in recent years is Wireless Smart Sensor Networks (WSSN). WSSN has a gateway which connects sensors to internetworks (having routers, switches, APs etc.) and the networked wireless sensors in WSSN are used to monitor and gather intelligence from the surrounding environment. In order to gather intelligence from surrounding environment WSSN makes use of Artificial Intelligence and Machine Learning techniques.

These WSSN networks includes modules for perceiving, assessing and communication. They can observe and retort to processes or incidences in a specified atmosphere. WSSN indicates the succeeding evolutionary growth phase in engineering like environmental control, industrial mechanization, traffic monitoring and robot control. WSSNs have numerous limited structures like power absorption, size, low price, scalability, suppleness and bounciness.

Artificial Intelligence (AI) is the imitation of human insight processes by computer systems. Recently, AI processes have attracted close contemplation of detection and have also been utilised effectively to elucidate problems in engineering. AI encompasses skilled systems, speech recognition, machine learning, deep learning platforms and robotic process automation. Machine Learning (ML) is primarily impelled as a special method for AI. ML can be delineated as the learning measures for augmentation of computer models that can extend the presentation of systems. Of late, usage of ML technologies in automation of health care systems has been skilled for identifying the diseases, forecasting the hazards of serious diseases and for monitoring patients with that diseases. Machine Learning also inspires many practical solutions that maximize resource utilization and prolong the lifespan of the network.

Compatibility and novelty are crucial factors to the health care domain. Hence it is required to develop suitable devices for health care by applying new techniques. For this, extensive research in AI is required. The evolution of ML techniques which are part of AI, is extremely growing. The application of ML techniques are extended to the fields of smart manufacturing, medical science, pharmacology, agriculture, archeology, games, business, and so forth.

The aim of this chapter is to provide inspiration and motivation to the Researchers by keeping track of new scientific accomplishments, to understand the tremendous potential of AI/ML in WSSN for health care applications.

2 Wireless Smart Sensor Networks (WSSN)

Smart environments depending on WSN has tremendous development in Engineering, such as industrial automation, video supervision, traffic monitoring and robotics. Sensors can be clustered into two varied classes: normal and smart. Normal sensors are those that necessitate steadfast exterior circuitry to do investigation on gestures, on fault recompense and filtering. If vast data is produced simultaneously, protecting it is also necessary, whereas smart sensors assimilate the sensor with the necessary buffers and acclimatising motherboard [1].

Wireless smart sensors network (WSSN) is a network that contains thousands of smart sensors. The smart sensors would be collecting data which will be directed for further processing [2]. Information provided by smart sensors which is required to be sent through a communication medium is characterized by a high-speed and bidirectional flow of data [3].

The IEEE 1451 standard for WSSN describes the formalized concept of smart sensor. The main idea of smart sensors comprises incorporation and preservation of disseminated sensor system to yield cost efficiency, smart and intellectual measurement, describing a common platform for calculating and interfacing several sensors of various sorts [4].

WSSN has two elementary aspects: the hardware and the software. The hardware arrangement entails a group that has smart sensor nodes, group head which acts as a processing component and expertise node that acts as an entry to a native zone network [1]. Figure 1 displays the hardware structure of a WSSN.

The software architecture as shown in Fig. 2 consists of the following components:

- Transducer Independent Interface (TII)
- Smart Transducer Interface Module (STIM)
- Transducer Electronic Data Sheet (TEDS)
- Communication protocol (CP).

WSSNs have various benefits like power consumption, magnitude, cost efficiency, scalability, suppleness, flexibility and hence plays an exceptional part in the field of remote observing [3]. Alternatively, the ecological observing is one of the important fields of application of this technology because of its features that permit the dimension of structures in diverse ecological surroundings such as crop management, protection of forest fires, agriculture, earthquakes, etc. It also utilises

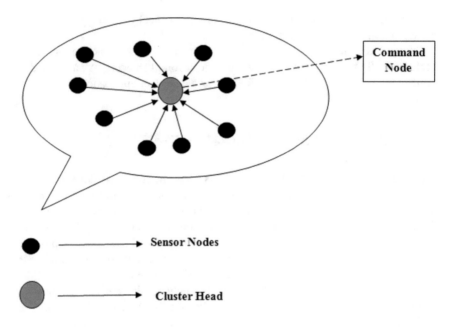

Fig. 1 Hardware structure of WSSN

Fig. 2 Software structure of
WSSN

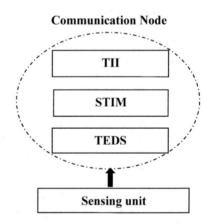

macro-instruments for assessing factors of landslides and environmental weather
casting [2].

Recent and probable use of WSN includes an extensive diversity of fields, which
has been conventionally perplexing to access owing to several causes comprising
probable destruction to humans, being at isolated places or being disseminated over
huge arenas, and being bound by severe geo or meteorological situations amongst
others.

The features of WSN are extremely huge number of sensor nodes, compressed disposition, varying topology assembly, restricted power source, calculation, storage and communication ability and lastly, the price. The applications and protocols running on WSSN are energy-efficient, scalable and robust. They also should familiarize to the varying atmosphere or context, and application choice and emphasis amid others. Several aspects comprising topography, climate and human-induced deliberate or non-intentional intrusion in the electromagnetic range will unfavourably disturb the positioned network and hence needing an excellent level of flexibility to varying conditions [5].

3 Brief Introduction to AI Techniques

AI is a novel method that investigates and progresses philosophies, approaches, technologies, and application systems for the replacement of human intellect. Usage of AI investigation is to expect that machines can do certain multifaceted responsibilities that need intellectual humans to carry out. That is, it is expected that the machine can substitute humans to resolve certain intricate challenges. In this procedure, it is not a monotonous mechanical action, but few needs human intelligence to partake in it [6].

AI has transformed information technology (IT). The novel frugality of IT has fashioned our life style. AI procedures have appealed close consideration of investigators and have also been used effectively to resolve difficulties in Engineering. On the other hand, for huge and intricate issues, AI procedures ingest substantial calculation time because of stochastic structure of the search methods. Hence, there is a possible necessity to improve effectual procedure to discover solutions under the restricted sources, time and cost in real world applications. This leads to the modern progresses in every single feature of AI technology together with Machine Learning (ML), Data Mining (DM) and Evolutionary Computation [7].

While the chief goal of AI is to progress systems that follow the intelligent and communication skills of a human being. The Distributed Artificial Intelligence (DAI) follows the same goal but concentrating on Human society. An example in present usage for the growth of DAI is according to the idea of multi-agent systems. A multi-agent structure is produced by a number of interrelating intellectual systems termed agents and can be employed as a software program, as a devoted computer or as a robot. Intelligent agents in a multi-agent system interrelate amongst one another to establish their arrangement, allocate responsibilities, and exchange knowledge. Notions associated to multi-agent systems, bogus people, and simulated societies, generate a novel and increasing pattern in calculations that includes problems as collaboration and competition, harmonization, association, communication and language etiquettes and communal intellect actions lead by agents [8].

3.1   Main Categories of AI Techniques

AI techniques have two main aspects: Machine learning (ML) and Deep learning (DL).

(i) **Machine Learning (ML)**

It is an application of AI that provides automatic learning from the earlier experiences, without any need of specific instructions. ML focuses on developing algorithms which can train and learn the patterns from the data. It makes optimum decisions and predictions based on the learned experience. The main goal of ML algorithms is to automate the process of training and learning without human assistance.

(ii) **Deep Learning (DL)**

A precise method of learning in ML is known as DL. It principally depends on procedures of neural networks. Currently, DL has created excessive development in arenas of image identification, speech identification, regular language processing, acoustic identification, social network filtering, machine translation and medical image investigation.

3.2   Neural Network Algorithms in ML and DL

(i) **Artificial Neural Networks (ANN)**

When AI is considered, we have to remark a renowned procedure in AI, known as Artificial Neural Networks (ANN). ANN is same as that of neural broadcast of the human brain, from one input component to the following input element to obtain an outcome. This is the rule of a standard neural network, which is to pretend the broadcast of data from nerves in the human brain. It handovers data from one neuron to other.

(ii) **Back Propagation Neural Network (BPNN)**

Later the discovery of the NN procedure, many issues have been resolved to some level. Simultaneously, people are continuously enhancing this procedure. Initially, an extensively utilised and very typical one is the BPNN which has additional concealed layer than the actual NN. There are additional hidden layers in the input layer and the output. It can significantly lessen the amount of evaluation and the trouble of evaluation by means of gradient decline.

(iii) **Convolutional Neural Network (CNN)**

However, once we have the BPNN, we realize that its computational load is still huge. At times, it flops to provide an ideal resolution within our adequate time series or it takes extension to provide the ideal resolution, which does not encounter the requirements of our certain applications. Then emanated the CNN, which is too a sort of NN procedure in core, but it enhances the content in the BPNN. This makes the calculation quicker and it addresses several issues. It advances the efficacy of its assessment by processing associated data simultaneously. It significantly decreases the computational intricacy amongst BPNNs. Hence, the CNN can presently attain the ideal resolution in a reckless time on several issues [9].

4 Brief Introduction to ML Techniques

The ML was initially propelled as an exclusive technique for AI in the late 1950s. Its focus is progressively shifted and also established further towards procedures that are computationally attainable and convincing over the years. The applications are established recently in several fields like spam discovery, bioinformatics, speech identification and fraud recognition. The embodiment of ML could be trapped by the subsequent two standard descriptions:

The learning processes for expansion of computer replicas that can augment system performance and provide procedures to the concern of information gaining.

Identifying and defining constancies and designs in training data by means of computational approaches that can enhance machine performance.

The present enhancements in ML and soft computing plans permit more valuable prediction replicas to be produced depending on a set of sizes. The cultured exemplary might be simply an elementary parametric function, educated from data, a pair of input variables-generally out-dated sizes or perspective, permitting output stage or adaptable to be anticipated exactly. The WSN's can encompass varied, several autonomous, economical and also a negligible power sensor nodes.

In this section, various ML methodologies and their learning etiquette are explained that will provide the basic knowledge for understanding the further units. Depending upon the learning methods, ML procedures have been characterized into supervised, unsupervised, semi-supervised and reinforcement [10]. Classification of ML procedures is given in Fig. 3.

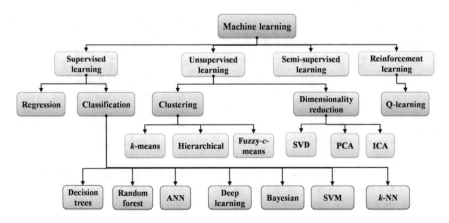

Fig. 3 Classification of ML procedures

4.1   Supervised Learning (SL)

SL is one of the most significant data analysis methodologies in ML. In SL, a collection of input and outputs (datasets with labels) are present which recognizes the connotation amid them when training the system.

- **Regression**
  Regression is a SL methodology which will envisage certain value (Y) depending on a specified set of structures (X). The variables in this model are incessant or measurable. This methodology estimates detailed conclusions with least imprecisions.
- **Decision Trees (DT)**
  DTs are a group of controlled ML methodology for cataloguing depending on a group of if-then guidelines to augment the readability. A DT comprises two sorts of nodes known as leaf nodes and decision nodes (option amongst alternatives). DT exploits a group by producing a training model based on decision rules determined from training data.
- **Random Forest (RF)**
  RF procedure is a SL method with an assembly of trees. Each tree in the forest provides a cataloguing. RF procedure functions in two phases, making of RF classifier and estimation of outcomes. RF functions proficiently for bigger datasets and mixed data.
- **Artificial Neural Networks (ANN)**
  An ANN is a SL method centred on the ideal of a human neuron for categorizing the data. ANN associated with an enormous number of neurons (processing units) that progress data and creates exact outcomes. ANN classically operates on layers which are associated with nodes and each node is linked with a dynamic function.
- **Deep Learning (DL)**
  DL is an organized ML methodology used for categorization, and it is a subsection of ANN. DL methodologies are the data learning design methods with multi-layer designs. It constitutes simple non-linear modules that vary the design from lower to upper layers to attain optimum resolution.
- **Support Vector Machine (SVM)**
  SVM is a SL classifier that identifies an ideal hyperplane to classify the data. SVM achieves the finest cataloguing by means of hyperplane and harmonized discrete opinion. Many training data is redundant as soon as a margin is documented and a collection of points assists to distinguish the margin. The points which are made use of to detect the margin are termed support vectors. SVM proposes the supreme arrangement from a proposed collection of data.
- **Bayesian**
  Bayesian is SL procedure centred on statistical learning methodologies. It identifies the associations amongst the datasets by learning the provisional individuality by means of numerous statistical approaches (example: Chi-square test).

- **K-Nearest Neighbor (KNN)**

  KNN is the most direct lethargic, SL technique in regression and classification. The training set takes into consideration an input from the feature space. KNN usually classifies depending on the distance amid fixed training samples and the trial sample. It utilizes several distance functions like Euclidean distance and Hamming distance.

4.2 Unsupervised Learning (UL)

In UL, there is no unlabelled data related to the inputs. Even the model attempt to excerpt the associations from the data. UL method is utilised for categorizing the group of same forms into groups, dimensionality lessening, and irregularity recognition from the data.

- **K-Means Clustering (KMC)**

  The KMC procedure effortlessly customs a firm number of groups from a provided dataset. At first, k numbers of arbitrary positions are taken into consideration and the other enduring points are related with the adjacent centers. As soon as the groups are created by covering all the points from the input, a novel centroid from every group is re-estimated.

- **Hierarchical Clustering (HC)**

  Hierarchical Clustering method clutches the same substances into groups that have a prearranged top-down or bottom-up command. Top-down hierarchical clustering is known as divisive clustering, where, a big sole barrier divides recursively till one group for every remark. Bottom-up graded grouping is known as agglomerative clustering, where every observation allocates to its group depending on thickness functions.

- **Fuzzy C-Means clustering (FCM)**

  FCM is also known as soft grouping established by Bezdek in 1981 by means of fuzzy set theory that allots the remark to one or more groups. In this method, groups are recognized depending on the resemblance dimensions like the power, distance or connectivity [10].

4.3 Semi-supervised Learning (SSL)

The SL procedures act on the labelled data and UL procedures act on unlabelled data, whereas the SSL procedures depends on the combination of both labelled and unlabelled data.

Two dissimilar zones are existing in SSL: (i) to determine the labels on unlabelled data (ii) to presume the labels on forthcoming evaluation data sets.

Depending on these points, SSL measures are fragmented into two groups: Transductive learning and inductive SSL. Transductive learning is applied to calculate the meticulous labels for a specified unlabelled dataset, while the inductive SSL acquires a function f: $X \mapsto Y$ so that it is likely to be a finest interpreter on future data.

SSL suits with numerous real-time uses like natural language processing, speech identification, spam filtering etc.

4.4   Reinforcement Learning (RL)

RL algorithm unceasingly study by intermingling with the atmosphere and collects data to take some actions. RL exploit the act by defining the ideal outcome from the atmosphere. Q-learning method is one of the model-free RL methods [10].

5   Scientific Applications of AI and Machine Learning (ML) Techniques

Some of the applications of ML are:

- For objective zone analysis issue and determining an ideal amount of sensor nodes to shield the zone.
- Energy-gathering offers a self-powered and enduring preservation for the WSNs positioned in the strict atmosphere. ML procedure progresses the act of WSNs to estimate the quantity of energy to be garnered within a specific time period.
- Sensor nodes may alter their position owing to certain inner or outside aspects. Precise localization is even and fast with the support of ML procedures.
- ML is utilised to separate the defective sensor nodes from usual sensor nodes and advance the effectiveness of the network.
- Routing data plays a chief part in augmenting the network lifespan. The active performance of sensor network needs active routing contrivances to augment the system act [10].

6   Need for Automation in WSSN

A smart sensor network has to robotically observe and uphold lighting, heating, and airing in the greenhouse. This sensor network has the following strategy purposes [11]:

Table 1 Sensor types used for automation applications

Sensors	HVAC	Lighting	Shading	Air quality and window control	Switching off devices	Standard HH uses	Security and safety
Temperature sensor	✓						
Moisture sensor				✓			
Optical sensors		✓	✓				
LDR sensor		✓					
Ph sensor			✓				
Water pump					✓		
Smoke and gas detectors				✓		✓	✓

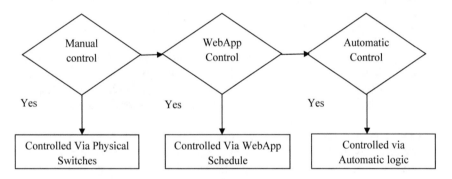

Fig. 4 Automation control architecture hierarchy

- Make the most of electrical energy effectiveness.
- Handle heating, cooling and aeration of the greenhouse.
- Utilise ancient and present weather data to evaluate future greenhouse energy requirements.
- Gather and submit greenhouse data (temperature, dampness, ambient light, etc.) to elementary school students and teachers via an available mobile or online interface.
- Permit for manual prevails of the whole automatic working in the greenhouse.

Table 1 given below tabulates the typical sensor types used for some of the automation applications.

The automation system enforces a hierarchy in its control architecture which is shown in Fig. 4.

7 Impact of ML Techniques in WSSN

The augmented quantity of sensors in devices will characteristically make huge data output, which postures a sombre task in handling and giving out the incredible quantity of sensual data. In addition, conventional processing methods in usual sensing devices are no more appropriate for methodically labelling, processing and examining the enthusiastic quantity of data.

The latest development of Machine Learning (ML) procedures and concepts has presented novel chances and vision to address these tasks sufficiently. The asset of ML procedures ability stands in its capability to acquire and systematize the abstraction of designs and structures from a specified group of data that conventionally needs a field expert to recognize.

The Smart Sensor System (SSS) has grabbed benefit of ML procedures and appropriate hardware to generate refined "smart" replicas which personalized precisely to identify modalities to obtain a more complete gratefulness of the system being checked.

ML technology looks very inspiring based on these descriptions for address concerns in WSNs as it allows applying old data to progress the efficacy of a network on accessible task, or even estimate the forth coming competence. For WSNs, using ML skill could be extremely good for many reasons like:

- Exceptional pursuing of active surroundings that adjust quickly with time. As a design, in soil tracking situation, it can be likely that the position of sensor nodes may alter due to soil scrape or ocean commotion and WSN based on machine learning can permit automated adjustment and cost-effective operation in certain sort of active surroundings.
- Contributing computationally and low-complexity mathematical replicas for composite atmospheres. In these surroundings, it is not easy to improve accurate mathematical replicas, and also problematic for sensor nodes to analyse the procedure suggestive of these kinds of arithmetical replicas. Under this kind of circumstances, WSN prejudiced by machine learning approaches which can offer low difficulty calculations for the system replicas, allowing its execution within sensor nodes.
- WSN centred on ML can be used for various application like lift automation and can be well used for CPS and m2m communication.
- As the WSN atmosphere is a benefit negligible, significant energy is scorched on projecting the theory with accuracy, and also for universal incident recognition sort situations, energy-efficiency, and forecast accuracy is fundamentally a compromise [12].

The ML techniques utilised in WSSN is classified into two broad categories namely Neural Network (NN) and non-Neural Network (non-NN) procedures. Non-NN procedures are further classified into:

(i)   Linear Regression (LR) which performs a regression task
(ii) Principal Components Analysis (PCA) which is a technique used to filter noisy datasets
(iii) Support Vector Machine (SVM) a supervised learning model that analyses data used for classification and regression analysis
(iv) Random Forest (RF) classifier that creates decision trees on data samples and then gets the prediction from each of them and finally selects the best solution by means of voting amongst others.

NN procedures are extremely effectual regarding aspect learning and abstraction. It needs fewer manual input when compared with non-Neural Network ML procedures. NN acquires intellectual structures from a specified dataset by the initiation of neuron nodes present inside the NN. Every single neuron is arithmetically activated by the enrichment data input which diffuse throughout the NN. The result of the NN is a consequence of dynamic neurons that are enthused by the dataset.

NN procedures are further classified into:

(i)   Long Short-Term Memory Network (LSTM) which is designed for sequence prediction problems.
(ii) Back propagation Neural Network (BPNN) which is used for training feed forward neural networks.
(iii) Recurrent neural Network (RNN) a class of artificial neural networks which processes variable length sequence of inputs
(iv) Convolutional Neural Network (CNN) commonly applied to analyze visual imagery have revealed foremost progression and developments in the field of medicine and engineering.

In common, non-NN procedures are more intricate to arrange and need more manual contribution to fine tune ML factors to attain application results. This can be accredited to the want for field awareness of the structures in a specified sensor dataset to precisely grow an application oriented non-NN smart model [13]. Figure 5 shows the Classification of ML techniques in WSSN.

8   Healthcare Applications of AI/ML in WSSN

The latest development of applications of AI in healthcare includes disease diagnostics, elderly people living assistance, biomedical information processing etc. It can be emphasized that, the application of AI in healthcare is still in its early stage. New progress and breakthroughs will continue to push the frontier and widen the scope of AI/ML application, and fast developments are eyed in the near future. Guoguang Rong et al. [14] has carried out a case study on filling of a dysfunctional urinary bladder.

Patients suffering from the storage and urination malfunctions of the bladder due to spinal cord injury or aging are provided with implantable neural stimulators for

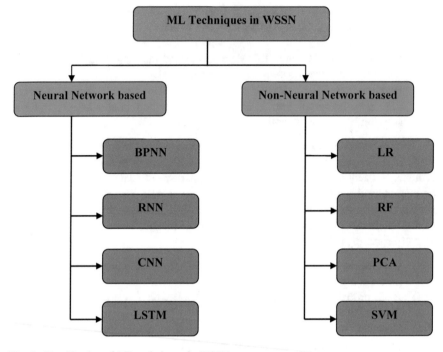

Fig. 5 Classification of ML techniques in WSSN

partial restoration of bladder function. In these cases a bladder sensor will detect the stored urine in the bladder. Based on the sensor input, the system will generate electrical stimulation to empty the bladder whenever it is full or at a regular interval for the patients with impaired sensations.

Using Digital signal processing we can sense the pressure in the urine through different neural activities found in the roots of the bladder where the changes that occur during the filling can also be detected.

Later the neural activity carried out by the sensors in the bladder is decoded by the proposed quantitative and qualitative methods of Machine Learning algorithms.

There have been numerous fruitful executions of ML smart sensor models with established ability of data processing and examination of an incredible quantity of data points. In this section, a summary of an extensive range of applied healthcare applications that assimilate ML procedures are explained.

Primary disease finding and diagnosis are the important criteria of interest for health care systems. The main sensors used in this system are Electronic Chemical Nose which aims at analysing and detecting diseases and Wearable Sensor to categorize mobility disorders through mobility patterns.

When united with ML procedures, electronic chemical nose is able to analyse these sorts of diseases in a non-invasive, suitable, and cost-effective way associated with those of traditional approaches. A SVMmodel can predict early symptoms of

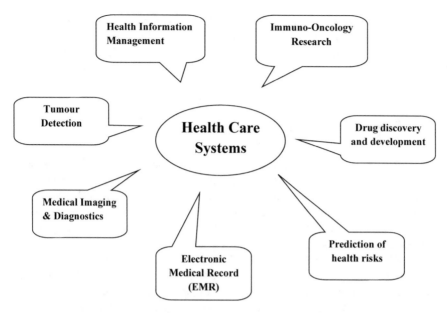

Fig. 6 Mind map of ML in health care applications

Alzheimer's disease by means of wearable sensors connected to the patients. A trained LSTM-RNN model is used to classify and predict diabetes for future glucose levels. Figure 6 shows the Mind Map of ML in health care applications.

8.1 Medical Image Diagnosis and Analysis

ML is employed to aid pathologists to do faster and more precise analysis of patients for novel kind of treatments or psychoanalyses.

Cancer cells lacking initial recognition can cause severe health problems that could cause death, and it is vital to perceive and analyse cancer cells at the very initial phases when it is curable. The constraint of traditional methods frequently ends in perceiving cancer at a much advanced stage. CNN procedures can be used to treat mammograms with the possibility to assist oncologists in initial breast cancer recognition that is not noticeable. This model uses Faster R-CNN framework to categorize malevolent, benevolent and ordinary tissues. The tumour boundary is segmented using CNN. This model shows significant improvement in prediction accuracy compared to radiologist diagnosis [15].

Biomedical images are imageries obtained from particular apparatus like Transmission Electron Microscope (TEM) or a Fluorescent Microscope (FM). These schemes are not restricted to CNN model that resolves cataloguing and division issues. They can be used to rebuild imageries obtained by a lesser determination image sensor. The CNN model was skilled once with 12,960 glossed micro particle images and positioned straight to the application. The result of CNN had confirmed the maximum possibility of cancer, where each tile score was distributed into RNN to categorize the whole slide image as malignant.

8.2 Prediction of Health Risks

ML is used to expect disease and aid doctors to treat and clients interfere prior, expect population fitness threat by recognizing forms and developing high threat indicators and model illness development and many more. Trained data for the chosen medical variables will be utilised by data mining classification procedures to discover the analysis styles. Next to the training procedure, these cataloging procedures utilize the learned analysis styles to expect the analysis for the given test data. RNN model of Deep learning has been used in real time diagnostic system in order to measure heart rate, blood pressure, body temperature and many other parameters using biomedical sensors. The Fuzzy logic decision model, Random Forest (RF), Q-learning are the other ML based approaches used for disease prediction [16].

8.3 Health Information Management

ML is utilised to augment fitness data management and interchange of fitness data, with the objective of updating workflows, enabling access to medical data and taming the precision and stream of fitness data. ML procedures are showing efficacy in creating interpretations about particular fitness threats and expecting fitness actions. For instance, neural network procedures have confirmed efficacy in perceiving strokes. Input variables examined by the procedure comprise stroke-related signs like paraesthesia of the arm or leg, severe misperception, image modification, issues with flexibility etc. This input data is examined to decide the possibility of stroke [17].

8.4 Drug Discovery and Development

Improvement of novel drugs is an onerous and expensive procedure. Certainly, so as to confirm both the patients' protection and drug efficacy, potential drugs must go through a modest and extended process.

ML and data science together with innovative laboratory technology are assisting to improve drugs with the aim of more rapidly curing patients at a lesser price. Because of the enormous quantity of biological and medical data obtainable nowadays, together with deep-rooted ML procedures, the scheme of principally automated drug improvement pipelines can now be seen.

This automation of the drug improvement procedure might be important to the present problem of less efficiency rate that medicinal enterprises now encounter. In ML procedure, feature selection goals at pruning or converting input data to only feed cherished data to the estimate model. For example, the profound DR method utilises an Auto-Encoder (AE) to produce useful structures from heterogenous drug-related data [18].

8.5 Tumour Detection

ML is used to discriminate amid tumors and healthy structure by means of 3D radiological images that support medical specialists in radiotherapy and surgical planning. The ML method can precisely categorize novel images into a set of images with tumors and a set of images with no tumor by recognising certain designs in the chosen feature values to discriminate amid the images of the two kinds [19].

A CNN technique has been used to detect brain tumors through brain Magnetic Resonance Imaging (MRI) images. While Extra Nodal Extension (ENE) of tumors in the head and neck cancer lymph nodes has been disreputably tough to analysis radiographically by doctors, a CNN-based model exhibited better than 85% exactness in recognizing this feature on diagnostic, contrast-enhanced CT scans. Other classification algorithms used for tumour detection are ANN, Tree J48., Navie Bayes and Lazy-IBk.

8.6 Electronic Medical Record (EMR)

Electronic Medical Record (EMR) is taken into consideration as an indispensable module of any healthcare association. Healthcare sources like doctors and nurses spend their valuable time in the course of their work gathering evidence from patients. Instances of the kinds of facts gathered are demographic evidence, therapeutic account and suggested medicine usage etc. EMRs are not suitable for high value data extraction since they are not scalable and are very costly to employ experts for data extraction [20].

AI is utilised to create prevailing EHR schemes more supple and intellectual. Taking medical notes with regular language processing permits doctors to concentrate on their patients instead of keyboards and screens. AI-supported tools assimilate with viable EHRs to help data gathering and medical note arrangement. Existing ML tools help in overcoming the difficulties in mining EMR to discover new approaches for drug development. With the help of ML, the cost of supporting EMR systems has been reduced by enhancing and regulating the mode those schemes are intended [21].

8.7 *Immuno-Oncology Research*

With the support of AI technology, ML is used for Immuno-Oncology investigation about how the physique's resistant system can battle cancer. Skilfully intended ML applications then have the possibility to support the human spectator in allocating biomarker scores to definite cell populations depending on morphological measures and staining features. ML procedures qualified on huge sample groups to discriminate PD-L1 positive immune cells (green) from tumor cell populations (red) signify an influential method for tissue cataloguing. Numerous ML methods like ANNs, Bayesian Networks (BNs), SVMs and Decision Trees (DTs) have been extensively utilised in cancer investigation for the growth of prognostic replicas which aids in effectual and precise decision making [22].

9   Conclusion

Wireless Smart Sensor Networks have gained a significant attention in healthcare domain with a wide range of capabilities. Application of WSSN in healthcare consists of wearable and implantable sensor nodes than can sense biological information and performs wireless transmission over a short distance. Machine learning provides a collection of techniques to enhance the ability of WSSN to adapt to the surrounding environment.

This chapter has provided a detailed study of ML and AI methods used in the automation process of various applications in WSSN such as Health care Systems.

10   Future Scope

Future work will focus on comparison of ML techniques in terms of various performance measures such as accuracy, execution time etc.

The skin cancer named Melanoma develops in the cells which controls the pigment in the skin. Treating patients with Advanced melanoma is a challenge since its complicated with immune related toxicity which results in secondary condition. Immune Checkpoint Inhibitors (ICI) which rely on biomarkers shows potential response in recent attempts [22]. The biomarkers are not scalable and requires more resources. Hence tissue histology features in melanoma can be identified prognostically using Machine Learning algorithms. Thus, our future work focuses on implementing various ML algorithms like DNN in the analysis of ICI response for patients with melanoma.

References

1. Al-Ali, R., Aji, Y.R., Othman, H.F., Fakhreddin, F.T.: Wireless smart sensors networks overview. IEEE (2005)
2. Chaudhari, M., Dharavath, S.: Study of smart sensors and their applications. Int. J. Adv. Res. Comput. Commun. Eng. **3**(1) (2014)
3. Technical Report, OECD code DSTI/ICCP/IE (2009)4/FINAL (2009)
4. Herrera-Quintero, L.F., Macia-Perez, F., Ramos-Morillo, H., Lago-Gonzalez, C.: Wireless smart sensors networks, systems, trends and its impact in environmental monitoring. IEEE (2009)
5. Serpen, G., Li, J., Liu, L.: AI-WSN: adaptive and intelligent wireless sensor network. Procedia Comput. Sci. **20**(2013), 406–413 (2013)
6. https://www.tutorialspoint.com/artificial_intelligence/artificial_intelligence_overview.htm
7. Zhang, Y., Balochian, S., Agarwal, P., Bhatnagar, V., Housheya, O.J.: Artificial intelligence and its applications. Math. Probl. Eng. **2014**(840491), 10 (2014)
8. Montoya, A., Restrepo, D.C., Ovalle, D.A.: Artificial intelligence for wireless sensor networks enhancement. In: Books, Smart Wireless Sensor Networks (2010). https://doi.org/10.5772/12962
9. https://medium.com/towards-artificial-intelligence/basic-concepts-of-artificial-intelligence-and-its-applications-294fb84bfc5e
10. Kumar, D.P., Amgoth, T., Annavarapu, C.S.R.: Machine learning algorithms for wireless sensor networks: a survey. Inf. Fusion **49**(2019), 1–25 (2019)
11. Meah, K., Forsyth, J., Moscola, J.: A smart sensor network for an automated urban greenhouse. In: International Conference on Robotics, Electrical and Signal Processing Techniques (ICREST) (2019)
12. Khan, Z.A., Samad, A.: A study of machine learning in wireless sensor network. Int. J. Comput. Netw. Appl. (IJCNA) **4**(4) (2017)
13. Ha, N., Xu, K., Ren, G., Mitchell, A., Ou, J.Z.: Machine learning-enabled smart sensor systems. Adv. Intell. Syst., 1–31 (2020)
14. Rong, G., Mendez, A., Assi, E.B., Zhao, B., Sawan, M.: Artificial intelligence in healthcare: review and prediction case studies. Engineering **6**(3), 291–301 (2020)
15. Saba, T.: Recent advancement in cancer detection using machine learning: systematic survey of decades, comparisons and challenges. J. Infect. Public Health **13**, 1274–1289 (2020)
16. Katake, K.: A study for heart disease prediction using IoT and deep learning. Int. J. Future Gener. Commun. Netw. **13**(3s), 1162–1168 (2020)
17. Jiang, F., Jiang, Y., Zhi, H., Dong, Y., Li, H., Ma, S.: Artificial intelligence in healthcare: past, present and future. Stroke Vasc. Neurol. **2**(4), 230–243 (2017)
18. Réda, C., Kaufmann, E., Delahaye-Duriez, A.: Machine learning applications in drug development. Computat. Struct. Biotechnol. J. **18**, 241–252 (2020)
19. Al-Ayyoub, M., Husari, G., Alabed-alaziz, A.: Machine learning approach for brain tumor detection (2015)
20. Alqahtani, A., Crowder, R., Wills, G.: Barriers to the adoption of EHR systems in the Kingdom of Saudi Arabia: an exploratory study using a systematic literature review. J. Health Inform. Dev. Countries **11**(2) (2017)
21. Tavares, J., Oliveira, T.: Electronic health record portal adoption: a cross country analysis. BMC Med. Inform. Decis. Mak. **17**(97), 2017 (2017)

22. Johannet, P., Coudray, N., Donnelly, D.M., Jour, G., Illa-Bochaca, I., Xia, Y., Johnson, D.B., Wheless, L., Patrinely, J.R., Nomikou, S., Rimm, D.L., Pavlick, A.C., Weber, J.S., Zhong, J., Tsirigos, A., Osman, I.: Using machine learning algorithms to predict immunotherapy response in patients with advanced melanoma. Clin Cancer Res Precis. Med. Imaging (2020). https://doi.org/10.1158/1078-0432.CCR-20-2415
23. Alsheikh, M.A., Lin, S., Niyato, D., Tan, H.-P.: Machine learning in wireless sensor networks: algorithms, strategies, and applications. arXiv:1405.4463v2 [cs.NI] 19 Mar 2015

Machine Learning Algorithms and Methodologies for Smart Sensor Networks

ML Algorithms for Smart Sensor Networks

Geetika Vashisht

Abstract Smart Sensor networks (SSNs) have surfaced as one of the most promising technologies of the Twenty First century by revolutionising the way in which the dynamic environments are monitored. The availability of economical micro sensors and the knowledge of machine learning has given a significant boost to the world of smart sensor networks. Machine Learning techniques can undoubtedly ameliorate SSNs. To adapt to the ever changing dynamic environments, smart sensor network exploits the machine learning approaches for optimum resource utilization and longevity of the network. Last decade has witnessed an over growing usage of machine learning algorithms in automating the SSNs. The paper reviews the machine learning algorithms employed in SSNs. The design issues are addressed, followed by the operational and the non-operational challenges that need immediate attention to push the technology further. The open issues that need to be investigated in future are addressed and future applications of SSNs are identified. This paper can be a guiding light for the new researchers entering the domain of SSNs by giving a detailed survey on the developments in the field in the last decade.

Keywords Smart sensor networks (SSN) · Wireless sensor networks (WSN) · ML (ML) · Localization · Outlier and anomaly detection · Fault detection · Compressive sensing · Data aggregation · Supervised techniques · Unsupervised techniques

G. Vashisht (✉)
Department of Computer Science, College of Vocational Studies,
University of Delhi, New Delhi 110017, India
e-mail: geetika.vashisht@cvs.du.ac.in

© The Author(s), under exclusive license to Springer Nature Switzerland AG 2022
U. Singh et al. (eds.), *Smart Sensor Networks*, Studies in Big Data 92,
https://doi.org/10.1007/978-3-030-77214-7_4

1   Introduction

Smart Sensor Networks is the most promising technologies for monitoring the real time environment that they sense using minimal resources. The ease of deploy-ability, size, lower cost of set up, flexibility, accuracy, reliability and scalability adds on to their rising demand over conventional networking solutions. To define the role of Smart Sensor Networks in a line, it can be said that they monitor and accumulate certain information from the desired field to send it for analysis by the sensor nodes at the base station. The overall goal of a smart sensor network depends upon the application but there are certain tasks that are common to the networks like gathering intelligence from the surrounding environment computing certain parameters, classifying the detected objects and then tracking objects.

With the advent of technologies and fields like IoT, ML, Networks and so on, the motivation to invent and automate everything feasible is also on a rise and thus giving a meteoric rise to smarter, cheaper and smaller wireless sensors. Usage of smart devices is exponentially increasing across the globe, making Smart Sensors Networks as one of the most researched area in the last decade.

1.1   Basics of Smart Sensor Networks

It comprises of a set of smart tiny devices aka sensor nodes connected together in a mesh topology, creating an invisible layer between the virtual and the physical world. These intelligent, compact and battery powered sensor nodes collaborate to communicate information obtained from the field under surveillance via wireless links to a base node which may process the data locally or can deliver to other networks like Internet (via a gateway) for distributed collaborative information processing. Figure 1 illustrates a typical SSN.

Smart sensors network deploys billions of sensor nodes and possibly at remote areas where minimum energy expenditure is needed and the automated need to reconfigure the network periodically is also crucial. At the same time sensors need to communicate among themselves especially in case of object detection to get a view of objects from all the angles and fuse the information obtained to conclude. The query raised shall be entertained by the network which can be done by a local sink node that can fuse the data from several nodes to come up with the response. Figure 2 highlights the features of the smart sensors networks just discussed.

Table 1 has the list of abbreviations used in this paper for readers' ease. The rest of the chapter is organized as follows. Section 2 reviews the existing machine learning algorithms briefly exploring their contribution to SSNs. Section 3 discusses ML solutions to address operational (essential to the basic functional aspect

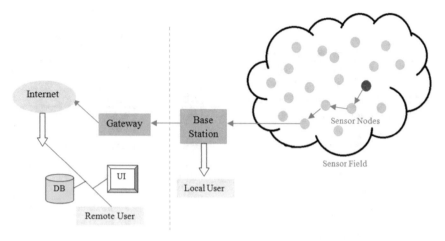

Fig. 1 A typical wireless sensor network

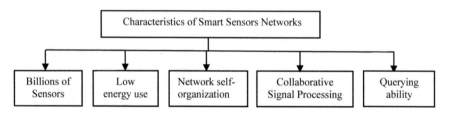

Fig. 2 Smart sensor networks' characteristics

of SSNs) and non-operational issues (the factors that determine non-functional requirements like quality or enhance performance of functional units) in SSNs. Section 4 briefs about the future applications of ML in smart sensor networks based on techniques, resources, clustering, coding etc. Finally, Sect. 5 concludes the chapter highlighting the comparative analysis of different ML algorithms in several SSNs applications.

2 Introduction to ML Approaches

ML approaches are categorised into supervised, unsupervised, semi-supervised and reinforcement learning. This section explores the available ML approaches and briefly highlights the contribution of each approach to the field of SSNs (Fig. 3).

Table 1 Abbreviations

AD	Anomaly detection	LLS	Linear least squares
ANN	Artificial neural network	MAC	Media access control
BIRCH	Balanced iterative reducing and clustering using hierarchies	ML	ML
CSMA	Carrier sense multiple access	NB	Naive Bayes
CURE	Clustering using representatives	PCA	Principal component analysis
DL	Deep learning	PSO	Particle swarm optimization
DT	Decision tree	QoS	Quality of service
EC	Evolutionary computation	RF	Random forest
FCM	Fussy c means clustering	RL	Reinforcement learning
FROMS	Feedback routing for optimizing multiple sinks in SSN	RLGR	RL based geographic routing
GPS	Global positioning system	SOM	Self organizing map
HA-A2L	High accuracy localization based on angle to landmark	SSN	Smart sensor networks
HC	Hierarchical clustering	SVD	Singular value decomposition
ICA	Independent component analysis	SVM	Support vector machine
IoT	Internet of things	SOM	Self organizing maps
kNN	k-nearest neighbour	SSN	Wireless sensor network
LWSVR	Light weight support vector regression	WLS-SVD	Weighted linear least squares based on SVD

2.1 Supervised Learning

It's a data processing approach to ML tasks that trains a model on some existing combination of input output pairs to predict the outcome for a new input [1]. Each example is a function f consisting of an input object, usually a vector x and the desired supervisory signal y (f: x → y). A supervised learning algorithm trains on the input set given and delivers the inferred function. For SSN, supervised learning algorithms can be extensively applied in fields such as localisation (e.g. [2–4]), anomaly detection (e.g. [5–7]), quality of service (e.g. [8, 9]), synchronisation (e.g. [10, 11]), security and intrusion detection (e.g. [12–14]) and data integrity (e.g. [15]).

DT—A decision tree (DT) is a supervised learning technique often called a classification tree or a reduction tree which maps target values from observations about an item. A decision tree chooses between decision nodes and gives the final outcome through leaf nodes [16]. Decision tree uses training data and applies classification rules to identify the outcome or class of the new incoming data by developing a training model. Researchers have been contributing considerably in addressing different design challenges in SSNs using decision trees such as

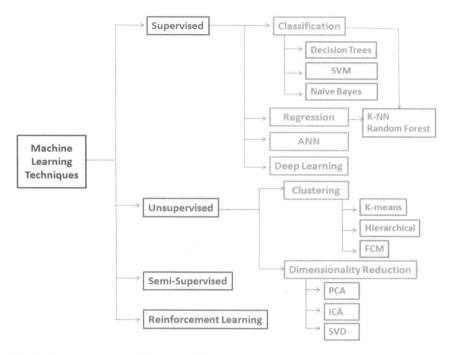

Fig. 3 Smart sensor networks' characteristics

identifying link reliability [17], connectivity [18], anomaly detection [19] and mobile sink path selection [20].

SVM—It solves two-group classification problems using classification algorithms. SVM model inputs a set of labelled training data for each category and then categorizes new text. The support vectors are formed by the boundary values and so it remains unaffected by the number of features that it comes across in the training data. SVM using coordinate individual observation and hyper plane performs the best classification [21]. This is the reason, SVM performs exceptionally well in cases with large number of features in comparison to training instances. SVM has been proved useful in various issues of SSN such as target tracking, connectivity problems [1, 18], routing [22], congestion control [23], fault detection [7, 24–26] and localisation [4, 27, 28].

NB—Naïve Bayes (NB) is a supervised learning technique with a statistical learning approach. Naive Bayes serves as a classification algorithm for binary (two-class) as well as multi-class classification problems. It is so called because it simplifies the calculation of the probabilities of each hypothesis based on a set of inputs so as to make its calculations traceable. One application of Bayesian inference in SSNs is assessing event consistency (θ) using incomplete data sets (D) by investigating prior knowledge about the environment [15]. A related statistical learning algorithm is Gaussian process regression (GPR) model [29]. It also has its application in synchronisation [10] and mobile sink path detection [30].

Regression—It is a supervised learning technique which produces target predicted value based on independent variables [31]. Depending on the type of relationship between dependent and independent variables, different regression models are used accordingly. To represent linear regression mathematically, Eq. 1 can be considered.

$$Y = f(x) + \varepsilon \tag{1}$$

Here, x is the independent variable i.e. input, Y is the dependent variable i.e. output and ε is used to represent possible random error. Regression deals with data aggregation [32], localisation [33] and energy harvesting [34] issues in SSNs.

RF—Random forest (RF) is a used for both classification as well as regression, but mainly used for classification problems [35]. RF, a supervised algorithm, works on the concept of ordinary forests like if they are made up of more trees it is said to be a more robust forest. In the similar fashion random forest algorithm by means of voting makes a prediction out of many decision trees on data samples by selecting the best solution. RF proves itself efficient in handling large and heterogeneous data. RF algorithm solves various issues in SSNs like coverage [36] and MAC protocol [37].

k-NN—The k-nearest neighbours (KNN) algorithm solves both classification and regression problems with a time complexity proportional to the size of the input dataset. As the size of data increases the prediction becomes slower. It's a supervised ML technique that works on the assumption that simular elements cohabitate. Several distance functions like the Euclidean are used by k-NN to compute the distance between two input points [38]. It finds the possible missing values from the feature space and also reduces the dimensionality [39, 40]. k-NN's simplicity makes it a best fit for query processing tasks in SSN. In SSNs, mobile sink, anomaly and fault detection [26, 41] and data aggregation [42] problems are solved using k-NN algorithm.

ANN—It works similar to a biological neuron network for classifying the data [43]. ANN, a supervised ML technique, is a collection of connected nodes or units called artificial neurons. Signals can be transmitted from one neuron to another like the synapses in a human brain. ANN typically operates on three layers—input layer, hidden layer(s) and output layers. It has high computation requirement but is still used widely in real time applications of SSN. ANN is popularly used to detect faulty sensor nodes [44], data aggregation [45] and congestion control [46].

DL—Deep Learning (DL) is based on ANN with representational learning. Information processing and communication patterns as in human nerve system is what inspires deep learning models. It mimics human brain to solve a problem and analyses the new data by comparing it with similar known objects in the existing database [47]. Chatbot, Alexa, Siri, self-driving cars are all gifts of DL. DL addresses several common issues in SSNs such as data quality estimation [12], anomaly and fault detection [7], energy harvesting [3] and routing [48].

2.2   Unsupervised Learning

Unsupervised Learning is an approach of ML which does not require any super-vision of model by the user. Instead, it allows the model to discover undetected patterns and information. Unsupervised learning mainly deals with the unlabelled information. The major contributions of this approach is in resolving various issues of SSNs such as routing [49–51], connectivity problem [52], data aggregation [53–56] and anomaly detection [57]. Clustering (hierarchical, k-means and fuzzy-c-means) and dimensionality reduction (PCA, ICA and SVD) branch out from unsupervised learning.

k-Means Clustering—It works iteratively using distance-based measurements to divide the given dataset into "k" distinct and pre-defined mutually exclusive clusters with high intra-cluster similarity and less inter-cluster similarity [58]. K-means clustering helps find optimal cluster heads for routing data towards the base station [50] in SSNs. It also helps find rendezvous points for mobile sink [59].

HC—Hierarchical clustering (HC) is also one of the unsupervised learning algorithm which groups the unlabelled data points together with similar charac-teristics. In Agglomerative hierarchical algorithms, every data point is considered a single cluster and then the pairs of clusters are successively merged or agglomerated (bottom-up approach). On the other side, in divisive hierarchical algorithms, the data points are treated as a single big cluster and after clustering one big cluster is divided into various small clusters (Top-down approach) [60]. SSNs issues like synchronization [61], data aggregation [62] and energy harvesting [63] are addressed efficiently by HC.

FCM Clustering—Also called as soft computing, FCM was developed by Bezdek in 1981. It uses fuzzy set theory to assign the observations to one or more clusters. Similar clusters on basis of intensity, distance or connectivity are identified by fuzzy c means. The algorithm iterates on the cluster until it finds an optimal cluster centre. FCM overpowers k means clustering for overlapped datasets. The time complexity which is usually higher than other approaches is highly dependent on several factors like the count of clusters, data points, dimensions and iterations. FCM technique solves several issues in SSNs like localization [64], connectivity [65] and mobile sink [66].

SVD—An unsupervised learning algorithm is widely used as a data reduction method in ML Routing issues of communicating data between sensor nodes are efficiently addressed by SVD [67]. Apart from that, data aggregation related issues are also addressed [68]. SVD also guarantees the optimal low-rank representation of the data [69].

PCA—A multivariate analysis feature extraction method for dimensionality reduction [70]. The main purpose of Principal Component Analysis (PCA) is to reduce the dimensions of a data set consisting of many variables correlated with each other, either lightly or heavily. Since SSNs' sensor nodes do continuous surveillance and huge volumes of data is transmitted between the sensor nodes, a need for dimensionality reduction crops up. PCA helps reduce dimensionality of data either at cluster head level or sensor level thereby reducing communication overheads. Several algorithms of SSNs such as localization [71], fault detection [1, 26] and target tracking [72] have adopted PCA.

ICA—A computational and statistical unsupervised technique to reveal factors that are hidden and underlie sets of random variables, signal or measurements. ICA is a generative model for multivariate data, typically given as a large sample database [73]. ICA analyse data from different fields such as business intelligence, digital images [74], psychometric measurements, web content and social networking [75].

2.3 Semi-supervised Learning

Semi-supervised learning majorly focuses on labeling the unlabeled data of the training set present and to predict the labels on future data sets for testing. Many real time applications of semi-supervised learning are observed in speech analysis, natural language processing, spam filtering, classification of web content, protein sequence classification etc. [76]. It also helps resolve localisation [77–80] and fault detection issues [81] for SSNs.

2.4 Reinforcement Learning (RL)

RL approach for SSNs involves learning by interaction with the environment with an award-reward mechanism. A benefit process is involved in this approach where a sensor node learns to seize ideal measures ensuring maximum advantage with experience [82]. SSNs routing problems are well addressed using reinforcement learning algorithms in which each one node tries to select measures which are anticipated to increase the extended benefits [83]. Here, the sensor node regularly updates the rewards it achieves based on the action it takes at a given state [84, 85]. "In Q-learning, each agent interact with environment and generate a sequence of observation as state action rewards (for example: $<s0, a0, r1, s1, a1, r2, s2, a, r3, ...>$)" [52, 86, 87].

3   ML in SSNs

Involvement of ML in automation of SSNs operations and addressing the issues in SSNs has gained a momentum. This is remarkable that there exist a ML technique to address the several operational and non-operational issues in SSNs. This section presents the classified summary of the work done by the eminent authors to deal with SSNs issues using ML.

3.1 Operational Challenges

Operational issues also referred to as functional issues are the ones that are essential for the basic operation of SSNs. Several operational or functional issues in design of SSNs can be resolved by consuming ML paradigms in the ways of SSNs work [88, 89].

Localization—Localization is about knowing the exact location of the sensor nodes, i.e. the geographic coordinates of the sensor nodes in a SSN. One seemingly simple way out is to use GPS hardware in every node of SSNs which is not an economical way and is not energy efficient [90]. At the same time, it can be infeasible to track the sensor nodes at remote locations and some specific indoor locations [89]. In this case, ML comes to rescue! Localization algorithms can be range-based, using distance or angle estimates for location estimations; or range-free, using connectivity information only between unknown nodes and nodes having its own location information aka landmark nodes [91]. Banihashemian and Adibnia [2] proposed a localization algorithm (LPSONN) that exploits PSO algorithm to optimize the neurons count in the neural network's layers resulting in less localization error rate and considerably low requirements for storing. Assaf et al. (2016) introduces a new range free localization algorithm that accounts for anisotropic signal attenuation using ANN. Gharghan et al. (2015) introduced hybrid PSO-ANN algorithm to improve the distance estimation accuracy of the mobile node. Shi et al. [92] and Bhatti [93] treats localization issue as a regression problem rather than a classification one. The approaches employing the combination of fuzzy-c means and SVM outperforms other ML approaches in terms of accuracy and time complexity [94] (Fig. 4).

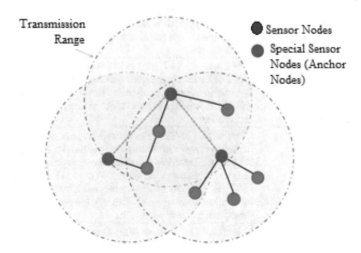

Fig. 4 Diagrammatic representation of localization in SSNs

Additionally, several other ML techniques that are used to address the localization issues are mentioned in Table 2.

Coverage and Connectivity—Coverage indicates how efficiently sensor nodes have been deployed in the area of interest. Connectivity, on the other hand, means that all the nodes are interacting with each other and to the sink node. ML approaches assist in identification of the minimum number of nodes required for covering the area of interest and in elimination of the faulty nodes; re-routing the traffic dynamically without data loss. In Fig. 5 two nodes are disconnected nodes (colored in red) and the white patch between the nodes indicates the coverage hole. Regression algorithms optimize network quality and reliability for centralized environments. RF-based algorithms come up as winner in improving the accuracy of the coverage test area when the number of features are randomly updated. NB is used to track human targets in the area under surveillance with considerably lesser computational complexity. k-means has been used for classification of the data of sensor nodes. RL-based algorithms improves network lifetime for both distributed and centralized networks.

ML techniques that are used to address the coverage and connectivity issues are mentioned in Table 3.

Table 2 Classified summary of the work done to address the localization issues in SSNs

SSN challenge	Analysis of ML approaches to address the localization issues	
Localization	Bayesian	Nguyen et al. (2016) uses Bayesian technique for localizing the unknown nodes location using posterior CramRao bound and sequential Monte Carlo to estimate source location and identifying unknown nodes respectively. Yu et al. [108] comes up with the lowest complexity approach that uses Bayesian approach to solve compressed sensing based localization. Both the former works are based on centralized environments
	Logic— vector PSO	Phoemphon et al. (2018) used fuzzy logic together with PSO for distributed environment (having static nodes) and a non-uniform node deployment. The time complexity of their algorithm is comparatively lower to other localization approaches
	SVM	Tang et al. (2017) comes up with an improved accuracy to develop a range-free localization method though complexity is slightly higher. SVM classifier discovers unknown nodes and transit matrix trains the system. Wang et al. [28] again uses SVM classifier to develop range-free localization method for centralized environment which is more energy-efficient as compared to other approaches
	DT	Almuzaini and Gulliver [109] proves that their novel range based algorithm based on DT classifier outperforms the approaches based on WLS-SVD and LLS

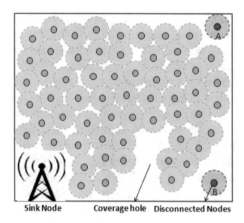

Fig. 5 Diagrammatic representation of coverage and connectivity in SSNs

Table 3 Classified summary of the work done in the area of coverage and connectivity algorithms in SSNs using ML approaches

SSN challenge	ML techniques used to address the connectivity ad coverage issues in SSNs	
Connectivity	Evolutionary computation	Xu et al. [110] experimentally proved that multi-objective evolutionary algorithm based on decomposition outperforms the existing algorithms in the literature on all the concerned parameters
	SVM	Kim et al. [111] presents a fully distributed SVM algorithm that improves the efficiency of connectivity and has a moderate communication complexity
	SVM + decision tree	Jian et al. [18] presents an energy efficient model for link quality estimation using smaller number of probe packets as compared to other techniques available
	Regression	Chang et al. [112] develops a regression based accuracy-aware interface design that measures the connectivity while reducing the overhead of the in-network aggregation approach
	RL	Ancillotti et al. (2017) presents a RL-probe for centralized environments that improves network life time by improving the link quality with a moderate complexity. Chen et al. [113] comes up with end-to-end data delivery reliability algorithm for centralized environments that improves network quality and reliability with a comparatively low complexity

Routing—Routing technique is required to establish communication between the sensor nodes and the base station. Since the resources are limited, the goal of routing protocols is to come up with energy efficient nodes enhancing the lifespan of the network. Routing protocols designed using ML approaches offer several benefits like optimal routing, enhanced lifespan of network, lesser energy depletion of nodes, lesser communication overhead and meet QoS requirements. Kadam and Srivastava (2012) performed routing in SSN to enhance network's life span and to transmit the information packages in the minimum span using RL. RL based routing algorithms can find the optimal routing path without the information about the global network structure but they need a long time to find the best route are making them not so suitable for highly dynamic environments. ML-Kernel-Linear regression algorithms contribute greatly in designing a distributed learning framework for SSNs because of their lesser overhead of the learning phase and good over-fitting results. SOM unsupervised learning contributes in detecting optimal routing paths based on multi-hop path QoS metrics but has higher overhead of the learning phase as compared to kernel-linear regression [95]. Q-MAP multicast, RLGR, Q-Probabilistic and FROMS are the latest routing protocols based on Q-learning algorithms [15]. ML-based routing protocols designs using RL, Regression, ANN, SVM, Bayesian, k-Means, SVD, SOM based approaches are pretty popular (Fig. 6).

Additionally, several other ML techniques that are used to address the routing issues are mentioned in Table 4.

MAC—Efficient MAC protocols can enhance the lifespan of the network. MAC protocols designed using ML approaches not only comes up with secure, economical, intelligent network with self-learning capability; it also reduces the end-to-end delay. The new nodes can easily reconfigure themselves. Kim and Park [96] proposed a CSMA contention based MAC protocol to manage the network utilizing the Bayesian ML algorithm to automate the channel allocation decision process. This avoids continuous sensing of data and saves network energy. Kulkarni and Venayagamoorthy [97] presents an application of neural networks in dealing with denial-of-service attacks. RL-MAC is an adaptive MAC protocol that uses RL algorithm for duty cycle management of the nodes in a network. RL algorithm is also used in novel ALOHA-QIR MAC protocol for dealing with

Fig. 6 Diagrammatic representation of routing in SSNs

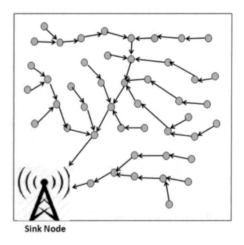

Sink Node

Table 4 Classified summary of the work done in the area of routing protocols in SSNs using ML approaches

SSN challenge	ML techniques to address the routing issues	
Routing	Bayesian	Arroyo-Valles et al. [114] came up with an energy-aware probabilistic decision model for intelligent routing
	Reinforcement learning	Ping and Ting [85] was first to introduce a routing strategy that adaptively learns and makes correct trade-offs. Experimentally proved that RL based approach outperforms the Q-learning based implementation [83]
	ANN	Turčaník [115] presented that neural networks for sensor routing table realization are faster irrespective of the size of the look-up table. It can adapt to the changing topology or environments

collisions [98]. Adaptive MAC layer protocols exploits DT classifiers [99]. MAC designs using RL, NN, RF, DT, Bayesian statistical models are pretty popular. ML techniques that are used to address the MAC issues are mentioned in Table 5.

Data Aggregation—It is the process of accumulating the data from the sensor nodes that has an affect over several parameters of SSNs viz. computational cost, energy efficiency, memory and communication overhead. ML approaches for data aggregation has significantly contributed in managing the energy of sensor nodes and keeping a check on the communication overhead [100]. PCA works really well

Table 5 Classified summary of the work done in the area of MAC algorithms in SSNs using ML approaches

SSN challenges	ML techniques to address MAC issues	
MAC	RF	Al-otaibi [37] experimentally proved that MAC address spoofing detection approach using RF outperformed traditional cluster based methods in terms of accuracy and prediction time
	RL	Rovcanin et al. [116] came up with a cooperative approach using RL for centralized as well as distributed SSNs with dynamic activation or deactivation policy for updating parameters of the system. The complexity is higher
		Phung et al. [117] presented a low complexity schedule-based approach based on RL that performed efficiently without overheads when a new node joins in the network. MAC layer protocol CSMA is used
		Kosunalp et al. [118] presented a low complexity contention-based approach based on RL that improvises the accuracy and channel performance by diminishing packet loss and adapting to the changes in the network. MAC layer protocol ALOHA is used
		Mustapha et al. [119] proposed a schedule-based dynamic approach based on RL with a moderate complexity to improvise the network lifespan channel allocation accuracy

in combination with two algorithms-Compressive Sensing and Expectation-maximization to enhance data aggregation. DT is popularly used to resolve the cluster head selection problem [101]. Regression techniques, k-NN, DT, PCA, Bayesian and genetic classifiers are used to improve the network lifetime. For reliable data transmission, Bayesian approach is preferred over other ML approaches. In order to reduce unnecessary transmission, Hierarchical clustering, PCA and SVD techniques contribute the most.

Additionally, several other ML techniques that are used to address the data aggregation issues are mentioned in Table 6.

Synchronization—Synchronization among the sensor nodes in SSNs is vital and ML efficiently ensures that the nodes are in synch with each other taking into account the changes in environment to resynchronize as and when required [11]. Pérez-Solano and Felici-Castell [10] used linear regression approach to propose a long-term synchronization method. Betta and Casinelli [102] proposed a model for low-cost SSNs applications. Additionally, several other ML techniques that are used to address the synchronization issues are mentioned in Table 7.

Congestion Control—Congestion happens in the channel when multiple nodes transmit simultaneously leading to collisions. Additionally, congestion can happen when the nodes' buffer that holds the packets to be sent out in the network overflows leading to packets lose! These problems causes delays as well as wastage of nodes' energy. The solution to the former is normally handled via MAC layer while the upper layers like transport or network layers are involved in the latter. Figure 7 illustrates the high packet arrival rate to a node that leads to node-level congestion whereas link-level congestion happens due to collision and lower bit transmission rate between two nodes.

ML techniques are pretty promising in dealing with these issues. RL and ANN approaches aid in traffic control increasing throughput. SVM improves the transmission rate while Fuzzy logic is used to keep a check on nodes' buffer size. Some other ML techniques that are used to address the congestion control issues are mentioned in Table 8.

Target Tracking—In target tracking, sensor nodes collaborate to monitor and report the positions of moving objects in the field of interest. Target tracking quality in SSNs face several challenges like node failure, energy consumption, and recovery mechanisms in case the target is lost. DL based approaches are efficient in tracking multiple targets in mobile SSNs. DT and SVM based algorithms contribute in classification of the targets in heterogeneous SSNs. Several taxonomies of target tracking algorithms are mentioned in Table 9.

Event Detection—It is the feature of detecting an event from the area under observation by continuously monitoring the environment to take certain decisions. False alarm rate and false detection rate must be less with high synchronization to successfully execute event detection. It is challenging because of limited resources like memory, power and computational resources. PCA and ICA based approaches help detect an event from complex forms of sensors data. EC and DL based approaches improves packet delivery ratio by achieving efficient duty cycles.

Table 6 Classified summary of the work done in the area of data aggregation in SSNs using ML approaches

SSN issues	ML techniques for data aggregation	
Data aggregation	k-NN	Li and Parker [120] proposed missing data estimation method for heterogeneous SSNs using k-NN. Though this approach has low complexity in finding out the temporal and spatial correlation between static sensor nodes, it is sensitive to the choice of initial seed value as well as outliers
	PCA	PCA based data aggregation approaches ace in reducing sensor nodes' energy consumption. High dimensional data is gracefully handled. The cluster heads compress the data received by the sensor nodes locally before forwarding it to the base station. The compressed data reduces the overhead of communication between the wireless sensor nodes which in turn increases energy efficiency of nodes, increases throughput and thus enhancing the networks' lifetime. The demerit is higher complexity and computational requirements [54, 55, 121–123]
	Decision tree	Edwards-murphy et al. [124] uses decision trees for data classification in heterogeneous SSNs achieving 95.38% accuracy. The complexity of the approach is moderate with an improved accuracy
	Regression	Atoui et al. [125] proposed a tree based data aggregation approach that reduces the communication overhead between nodes or cluster heads to base stations. Additionally they proved that Euclidean distance is not an optimal measure since a spatial and temporal correlation between the nodes has to be determined as well. Gispan et al. [126] proposes a data aggregation approach for decentralized network using linear regression model to evaluate parameters for accumulating information
	SVD	The approach based on SVD to detect the missing information in an image is accurate only if the number of sensor nodes in a network are more [68]
	ANN	Neural networks resolves clustering problem by coming up with self-managed clusters [46]
	Bayesian	Das [127] used Bayesian approach to develop multi-sensor data fusion mechanism to gather the data. This approach adapts to the dynamic nature of the network. Hwang et al. [128] presented a lower complexity Bayesian based compressive sensing approach outperforming the traditional ones. Decrease active sensor nodes and estimation error while aggregating the data from the sensor nodes. Wang and Bertino [129] proposed an approach where each sensor node compresses data before sending to the cluster heads and uses arithmetic coding to locate the non-traceable packet paths

Table 7 Classified summary of the work done to address the issues related to synchronization in SSNs

SSN issues	ML approaches for synchronization in SSNs	
Synchronization	Regression	Pérez-Solano and Felici-Castell [10] proposed a highly accurate and efficient synchronization method to adapt to the dynamic nature of SSNs by re-synchronizing to deal with the change in the clock drift between sensor nodes
		Betta and Casinelli [102] resolves synchronization issues using regression; stressing mainly on clock wander, jitter, latency, clock drift and resolution of clock
	Bayesian + regression	Pérez-Solano and Felici-Castell [10] proposed a synchronization method using linear regression and Bayesian to give higher accuracy rate and lesser synchronization error as compared to traditional approaches

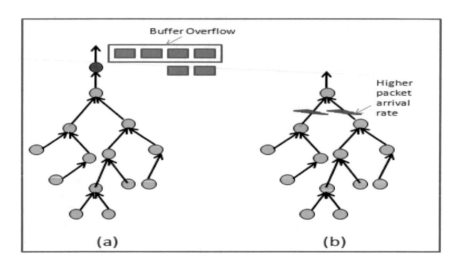

Fig. 7 a Node level congestion in SSNs **b** link level congestion in SSNs

Additionally, several ML techniques that are used to detect events mentioned in Table 10.

Mobile Sink—Energy hole problem can be addressed by mobile sink that visits each sensor node to collect information directly but scheduling it in an efficient delay-aware manner is an issue. To resolve this, mobile sink selectively visit a few nodes in SSNs called rendezvous points. Rest of the nodes send their data to nearest rendezvous points [100]. RF and RL based algorithms contributes in identifying the optimal rendezvous points. EC based algorithms help to select the optimal rendezvous points and an optimal mobile sink path to sensor nodes. Also, RF enables

Table 8 Classified summary of the work done to address the issues related to congestion control in SSNs

SSN issues	ML approaches to address the issues related to congestion control in SSNs	
Congestion control	ANN	Alsheikh et al. [130] uses neural for traffic control. It reduces energy consumption by nodes by transmitting compressed data. Minimized end-to-end delay is there between the nodes. Continuous data flow is achieved using hop-by-hop control pattern. Rezaee [131] develops an approach for minimizing packet loss ratio using efficient queue management
	Fuzzy logic	Betta and Casinelli [102] proposed synchronization method based on SVM for low-cost SSNs' applications by focusing on jitter, latency, clock drift, clock wander and resolution of clock
	SVM	SVM outperforms in terms of accuracy [23]. With the change in traffic, nodes' transmission rates are adjusted leading to efficient energy consumption, increased throughput and reduced latency

Table 9 Classified summary of the work done in the area of target tracking in SSNs using ML approaches

SSN issues	ML techniques for target tracking in SSNs	
Target tracking	Bayesian	Braca et al. [132] exploited Bayesian to develop an accurate target tracking together with data fusion
	Bayesian + RL	Das et al. [133] came up with an accurate energy efficient approach that used Bayesian network for monitoring events and RL to identify the duration of sleep schedules of the sensor nodes
	PCA	Oikonomou et al. [72] exploits PCA for processing the information from the field under observation using wireless sensing. Nodes are static and the network lifetime improves considerably
	Q-learning	Wei et al. [134] proposed a target tracking model for efficient task scheduling for cooperative sensor nodes

the identification of data forwarding routes and delay aware routing of mobile ML techniques that are used to address the mobile sink related issues are mentioned in Table 11.

Energy Harvesting—The lifetime of the network is directly proportional to the battery power of the sensor nodes arising the need of energy harvesting approaches for sensor nodes. Energy, a well-known issue in SSNs, aims at coming up with energy efficient and self-powered sensor nodes for an uninterrupted work. ML-based energy harvesting approaches help forecast the amount of energy that needs to be harvested in a particular time span [103] and reduces the computational complexity in calculating the amount of energy harvested. Shaikh and Zeadally [104] categorizes energy harvesting techniques in two categories-ambient and external sources. Radio-frequency based energy harvesting, solar-based energy

Table 10 Classified summary of the work done in the area of event detection in SSNs using ML approaches

SSN issues	ML approaches for event detection in SSNs	
Event detection	Bayesian	Avci et al. [135] proposed Bayesian learning-based motion detection method for detection of multiple objects in motion to reduce latency and efficient energy consumption
	RL	Ye and Zhang [136] presented sleep/wake-up scheduling approach using RL that enhances network lifetime with high packet delivery ratio though with a delay in packets delivery
	Deep learning + PCA	Wu et al. [137] 225 concludes that deep learning contributes to efficient duty cycle management. PCA does dimensionality reduction while deep learning approach is used for training the model. The approach has a higher accuracy of 94.12%

Table 11 Classified summary of the work done in the area of mobile sink in SSNs using ML approaches

SSN issues	ML approaches for mobile sink in SSNs	
Mobile sink	Hierarchical clustering	Zhang et al. [138] presented a node-density-based-clustering and mobile collection approach that gives increased throughput, reduced latency and an improved network lifetime
	NB	A method of data collection from sensor nodes using a mobile sink based on NB classifier efficiently outperforms traditional approaches Kim and Kim [139]
	Fuzzy clustering approach	Banimelhem and Abu-hantash [140] uses fuzzy logic to recalculate the location of mobile sink in every iteration based on the nodes' remaining energy. This approach outperforms other approaches in terms of energy efficiency and the count of active nodes in the network

harvesting, thermal-based energy harvesting and flow-based energy harvesting are the ambient sources while external sources can be mechanical-based or human-based energy harvesting. Though the harvesting capabilities of each of the sources mentioned is unique and depends largely on underlying hardware, the current state-of-the-art model is still immature. ML techniques that are used to address the energy harvesting issues are mentioned in Table 12.

3.2 Non-operational Challenges

Non-operational aspects of SSNs, also referred to as non-functional, are the ones associated with evaluating the quality and efficiency of operational components of

Table 12 Classified summary of the work done in the area of energy harvesting algorithms in SSNs using ML approaches

SSN issues	ML approaches to address the issues in the area of energy harvesting algorithms in SSNs	
Energy harvesting	Q-learning	Kosunalp [103] concluded that Q-learning based solar energy prediction approach for centralized environments aces in forecasting the amount of energy to be harvested in a frame of time by learning from experiences with a moderate complexity
	Hierarchical clustering	Awan and Saleem [63] used hierarchical clustering approach for distributed environments. Optimal locations for static cluster heads that use renewable sources (solar/wind) is found out in this approach achieving lesser power consumption by non-renewable sensor nodes with a low complexity
	RL	Hsu et al. [141] proposed a RL based approach that outperforms the traditional approaches to automatically adjust duty cycles
		Aoudia et al. [142] presented RL based solar energy management approach to balance energy consumptions of the sensor nodes and to maintain energy harvesting for centralized SSNs. It comes up with a comparatively lower complexity

SSNs, for instance, security and integrity of data, QoS and fault tolerance [89]. Prominent work in this area is discussed next.

Outlier and Anomaly Detection—ML approaches can be exploited to detect malicious activities, vulnerabilities, intrusions and potential threats in SSNs so as to increase the lifespan of the SSN ensuring the reliability at the same time. Outlier detection in live streaming environment is challenging because of resource constraints like finite memory and bandwidth available to nodes. Bayesian belief networks identifies the conditional dependencies among nodes to discover potential outliers. Non-parametric k-NN-based algorithm though requires large memory, can efficiently be used to for outlier detection and to find the data of missing nodes by computing the average value of the 'k'-nearest neighbors. SVM has been implemented in detecting selective forwarding and black hole attacks [14]. SOM-based algorithms are also capable of analyzing the attacks in the network but they don't function well in large and complex data sets. Additionally, several other ML techniques that are used to cater to the Outlier and Anomaly Detection are mentioned in Table 13.

Fault Detection—Fault Detection is a necessity for the modern SSNs. Traditionally, fault tolerance was dealt with by introducing additional sensor nodes as backup but that added to complexity and cost of the network. Intelligent fault detection methods can be deployed by ML approaches by comparing the actual network model with the faulty one to find the fault or to obtain information from the neighboring nodes. Moustapha and Selmic [105] proposed a non-static fault detection model. Warriach and Tei (2017) experimentally proved that kNN

Table 13 Classified summary of the work done for outlier and anomaly detection in SSNs

SSN challenge	ML approaches to address the issues related to outlier and anomaly detection in SSNs	
Anomaly and fault detection	k-means	Wazid and Das [143] proposed a moderate complexity model for centralized environment to detect hybrid attacks using k-means to achieve improved accuracy in detecting malicious nodes automatically
	SVM	Zidi et al. [7] proves the efficacy of SVM in fault detection (negative alerts). This approach uses a kernel function to serve the purpose with a relatively low complexity and 99% accuracy rate
	SVM + DBSCAN	Saeedi Emadi and Mazinani [5] presented an intrusion detection approach for centralized environments that exploits the classification capabilities of SVM to improvise DBSCAN's DoS attacks detection accuracy. The complexity is slightly higher than other approaches
	Decision tree	Garofalo et al. [19] proves that decision trees classification obtains higher intrusion detection rate in distributed environments with a moderate complexity

Table 14 Classified summary of the work done for fault detection in SSNs

SSN challenge	ML approaches to address the issues related to fault detection in SSNs	
Fault detection	Bayesian	Meng et al. [144] uses Bayesian inference to identify insider attacks via malicious nodes. Titouna et al. [145] proves that Bayesian classifier efficiently identifies faulty sensor nodes
	SVM	Jiang et al. [25] proves that SVM model based on the cuckoo search algorithm outperforms traditional ways to predict dynamic measurement errors for sensors
	SVM + PCA	Cheng et al. [146] proves that distributed fault detection mechanism using support vector regression reduces communication to sensor nodes with a low false alarm rate and higher accuracy. Islam et al. [24] uses multi-class SVM classifier for training and PCA to extract main features. The approach comes up with 99.80% accuracy which is quite greater than the traditional approach
	Fuzzy logic	Chanak and Banerjee [44] proves that fuzzy logic based faulty nodes classification comes up better than traditional methods. A data routing algorithm is proposed that reuses faulty nodes and improves QoS throughout the network lifetime

outperformed SVM and NB ML algorithms in obtaining a better fault detection rate on given performance metrics. Additionally, several other ML techniques that are used to cater to the fault detection in SSNs are mentioned in Table 14.

Quality of Service (QoS)—QoS defines the level of service given to the users by SSNs and is categorized as network-specific (bandwidth and nodes' energy consumption parameters) and application specific parameters (number of active sensor nodes). The high-priority delivery of the real-time data is a biggest challenge to maintain the QoS for SSNs, with resource management, unbalanced traffic, data redundancy and scalability being the additional challenges. ML approaches contribute in dealing with the challenges of QoS. ML approaches can be used to automatically recognize different types of streams Khan and Samad [89], NN for QoS estimation, RL for constraint satisfied service composition and cross layer communication framework, ANN for faulty node detection and data fusion and energy balancing [100]. Additional ML techniques for QoS in SSNs are shown in Table 15.

4 Future Applications of ML in SSNs

As SSNs continue to grow, the need for novel techniques to come up with energy efficient nodes keeping in check the parameters like cost, delay traffic and lifespan of the network is the need of the hour [106]. There are still many open issues in SSNs that need further research. ML can be applied to resolve these issues.

Compressive Sensing and Sparse Coding—Data compression and dimensionality reduction are the important considerations in data transmission. As stated by Kimura and Latifi [107], data transmission consumes eighty percent of the nodes' energy, cropping up network management and communication related design challenges. ML techniques can efficiently overcome the high computational and memory needs of traditional data compression techniques [46].

Hierarchical Clustering to Detect Data Spatial and Temporal Correlations —As described in Sect. 3, hierarchical clustering is an unsupervised ML technique

Table 15 Classified summary of the work done for QoS in SSNs

SSN challenge	ML approaches to address the issues related to QoS in SSNs	
QoS	NN	Yuan and Member [147] proves the efficacy of neural network based link quality estimation algorithm where signal-to-noise ratio is used as the link quality metric
	RL	Chandrakala [148] uses multi-agent RL for effective self-configuration and self-optimization and proves that a better QoS, throughput and end-to-end delay
	Fuzzy logic	Collotta et al. [149] provides a fuzzy-based solution for data fusion to obtain better QoS. Each sensor node has a fuzzy logic controller that takes several parameters as input and determines the probability of occurrence of an event. The data is forwarded to base station only if the probability estimated by the fuzzy system is higher than a pre-defined threshold

that can exploit clustering criteria like spatial and temporal correlations of set of objects to cluster them. This technique has potential for providing energy efficient solutions. BIRCH and CURE are typical methods of hierarchical clustering [15].

Using Resource Management Using ML—Energy Efficient SSNs can be designed by identifying the activities that dramatically drain the nodes' energy and by incorporating the efficient communication related protocols that includes physical, MAC and networking layer protocols. ML techniques can enable the nodes to self-organize themselves by optimally managing their resources and energy [15].

Distributed and Adaptive ML Techniques for SSNs—Distributed ML techniques best suit the design of SSNs where the nodes avoid the heavy computational tasks but self-organize to adapt to current environment situations and behave accordingly. The distributed learning techniques need not focus on the information of the entire network rather the focus is on the individual nodes requiring lesser memory and computational power [15].

5 Concluding Remarks

The research in the field of SSNs is immensely dynamic and progressing at a noticeable pace. This paper aimed to cover the important aspects of the field and the classified work in the last decade. ML algorithms are briefly covered for readers' convenience. The stumbling blocks in SSNs are identified and are addressed by ML techniques such as localization, anomaly detection, fault nodes detection, routing, data aggregation, MAC protocols, synchronization, congestion control, energy harvesting and mobile sink path determination. Apart from the operational issues, non-operational challenges that should be addressed in order to push the technology further are identified.

Future applications of ML in the ever growing field of SSNs presented in the chapter can encourage new research. The classified summary of the work done in the last decade presented in this paper together with a rich bibliography content gives a valuable insight into the field of SSNs.

References

1. Drozdov, V.N., Kim, V.A., Lazebnik, L.B.: Modern approach to the prevention and treatment of NSAID-gastropathy. In: Éksperimental'naia i klinicheskaia gastroénterologiia = Exp. Clin. Gastroenterol. 2 (2011)
2. Banihashemian, S.S., Adibnia, F., Sarram, M.A.: A new range-free and storage-efficient localization algorithm using neural networks in wireless sensor networks. Wireless Pers. Commun. **98**(1), 1547–1568 (2018). https://doi.org/10.1007/s11277-017-4934-4
3. Lu, C.H., Fu, L.C.: Robust location-aware activity recognition using wireless sensor network in an attentive home. IEEE Trans. Autom. Sci. Eng. **6**(4), 598–609 (2009). https://doi.org/10.1109/TASE.2009.2021981

4. Shareef, A., Zhu, Y., Musavi, M.: Localization Using Neural Networks in Wireless Sensor Networks (2009). https://doi.org/10.4108/icst.mobilware2008.2901
5. Saeedi Emadi, H., Mazinani, S.M.: A novel anomaly detection algorithm using DBSCAN and SVM in wireless sensor networks. Wireless Pers. Commun. 98(2), 2025–2035 (2018). https://doi.org/10.1007/s11277-017-4961-1
6. Xie, M., Hu, J., Han, S., Chen, H.H.: Scalable hypergrid k-NN-based online anomaly detection in wireless sensor networks. IEEE Trans. Parallel Distrib. Syst. 24(8), 1661–1670 (2013). https://doi.org/10.1109/TPDS.2012.261
7. Zidi, S., Moulahi, T., Alaya, B.: Fault detection in wireless sensor networks through SVM classifier. IEEE Sens. J. 18(1), 340–347 (2018). https://doi.org/10.1109/JSEN.2017.2771226
8. Anuradha, Solanki, A.K., Kumar, H., Singh, K.K.: Calculation and evaluation of network reliability using ANN approach. Procedia Comput. Sci. 167(2019), 2153–2163 (2020). https://doi.org/10.1016/j.procs.2020.03.265
9. Wang, Y., Martonosi, M., Peh, L.-S.: Predicting link quality using supervised learning in wireless sensor networks. ACM SIGMOBILE Mob. Comput. Commun. Rev. 11(3), 71–83 (2007). https://doi.org/10.1145/1317425.1317434
10. Pérez-Solano, J.J., Felici-Castell, S.: Improving time synchronization in wireless sensor networks using Bayesian inference. J. Netw. Comput. Appl. 82, 47–55 (2017). https://doi.org/10.1016/j.jnca.2017.01.007
11. Capriglione, D., Casinelli, D., Ferrigno, L.: Analysis of quantities influencing the performance of time synchronization based on linear regression in low cost WSNs. Meas. J. Int. Meas. Confederation 77, 105–116 (2016). https://doi.org/10.1016/j.measurement.2015.08.039
12. Janakiram, D., Reddy V.A.M., Kumar, A.V.U.P.: Outlier detection in wireless sensor networks using bayesian belief networks. In: First International Conference on Communication System Software and Middleware, Comsware (2006). https://doi.org/10.1109/comswa.2006.1665221
13. Branch, J.W., Giannella, C., Szymanski, B., Wolff, R., Kargupta, H.: In-network outlier detection in wireless sensor networks. In: Knowledge and Information Systems, vol. 34, issue 1 (2013). https://doi.org/10.1007/s10115-011-0474-5
14. Kaplantzis, S., Shilton, A., Mani, N., Şekerciğlu, Y.A.: Detecting selective forwarding attacks in wireless sensor networks using support vector machines. In: Proceedings of the 2007 International Conference on Intelligent Sensors, Sensor Networks and Information Processing, ISSNIP, pp. 335–340 (2007). https://doi.org/10.1109/ISSNIP.2007.4496866
15. Alsheikh, M.A., Lin, S., Niyato, D., Tan, H.P.: Machine learning in wireless sensor networks: algorithms, strategies, and applications. IEEE Commun. Surv. Tutorials 16(4), 1996–2018 (2014). https://doi.org/10.1109/COMST.2014.2320099
16. Quinlan, J.R.: Induction of decision trees. Mach. Learn. 1(1), 81–106 (1986). https://doi.org/10.1007/bf00116251
17. Tseng, C., Chen, C., Lin, T., Wu, Y., Lin, C., Lin, S., Liao, C., Szu, S., Yen, C., Lin, K., Wu, Z., Examiner, P., Lee, A., Steven, M., Palmer, E.A.: United States Patent, vol. 2, issue 12 (2010)
18. Jian, S., Song, L., Linlan, L., Liqin, Z., Gang, H.: Research on link quality estimation mechanism for wireless sensor networks based on support vector machine. Chin. J. Electron. 26(2), 377–384 (2017). https://doi.org/10.1049/cje.2017.01.013
19. Garofalo, A., Di Sarno, C., Formicola, V.: Enhancing intrusion detection in wireless sensor networks through decision trees. Lect. Notes Comput. Sci. (Including Subseries Lect. Notes Artif. Intell. Lect. Notes Bioinform.) 7869, 1–15 (2013). https://doi.org/10.1007/978-3-642-38789-0_1
20. Kim, S.D., Lee, E., Choi, W.: Newton's algorithm for magnetohydrodynamic equations with the initial guess from Stokes-like problem. J. Comput. Appl. Math. 309, 1–10 (2017). https://doi.org/10.1016/j.cam.2016.06.022
21. Vapnik, V.N.: An overview of statistical learning theory. IEEE Trans. Neural Netw. 10(5), 988–999 (1999). https://doi.org/10.1109/72.788640

22. Khan, F.A., Yousaf, A., Reindl, L.M.: Using capacitive glocal technique. Eur. Freq. Time Forum (EFTF) **2016**, 1–4 (2016). https://doi.org/10.1109/EFTF.2016.7477836
23. Gholipour, M., Haghighat, A.T., Meybodi, M.R.: Hop-by-Hop congestion avoidance in wireless sensor networks based on genetic support vector machine. Neurocomputing **223**, 63–76 (2017). https://doi.org/10.1016/j.neucom.2016.10.035
24. Islam, M.R., Uddin, J., Kim, J.M.: Acoustic emission sensor network based fault diagnosis of induction motors using a gabor filter and multiclass support vector machines. Ad-Hoc Sensor Wireless Netw. **34**(1–4), 273–287 (2016)
25. Jiang, M., Luo, J., Jiang, D., Xiong, J., Song, H., Shen, J.: A cuckoo search-support vector machine model for predicting dynamic measurement errors of sensors. IEEE Access **4**(c), 5030–5037 (2016). https://doi.org/10.1109/ACCESS.2016.2605041
26. Sun, Q.Y, Sun, Y.M., Liu, X.J., Xie, Y.X., Chen, X.G.: Study on fault diagnosis algorithm in WSN nodes based on RPCA model and SVDD for multi-class classification. Cluster Comput. **22**, 6043–6057 (2019). https://doi.org/10.1007/s10586-018-1793-z
27. Hong, J., Ohtsuki, T.: Signal eigenvector-based device-free passive localization using array sensor. IEEE Trans. Veh. Technol. **64**(4), 1354–1363 (2015). https://doi.org/10.1109/TVT.2015.2397436
28. Wang, Z., Zhang, H., Lu, T., Sun, Y., Liu, X.: A new range-free localisation in wireless sensor networks using support vector machine. Int. J. Electron. **105**(2), 244–261 (2018). https://doi.org/10.1080/00207217.2017.1357198
29. Shionoya, S., Ban, I., Nakata, Y., Matsubara, J., Hirai, M., Kawai, S.: Involvement of the iliac artery in Buerger's disease (pathogenesis and arterial reconstruction). J. Cardiovasc. Surg. **19**(1), 69–76 (1978)
30. Tashtarian, F., Yaghmaee Moghaddam, M.H., Sohraby, K., Effati, S.: ODT: optimal deadline-based trajectory for mobile sinks in WSN: a decision tree and dynamic programming approach. Comput. Netw. **77**(December), 128–143 (2015). https://doi.org/10.1016/j.comnet.2014.12.003
31. Doan, T., Kalita, J.: Selecting machine learning algorithms using regression models. In: Proceedings—15th IEEE International Conference on Data Mining Workshop, ICDMW 2015, pp. 1498–1505 (2016). https://doi.org/10.1109/ICDMW.2015.43
32. Zahara, S.I., Ilyas, M., Zia, T.: A study of comparative analysis of regression algorithms for reusability evaluation of object oriented based software components. In: ICOSST 2013— 2013 International Conference on Open Source Systems and Technologies, Proceedings, pp. 75–80 (2013). https://doi.org/10.1109/ICOSST.2013.6720609
33. Zhao, W., Su, S., Shao, F.: Improved DV-hop algorithm using locally weighted linear regression in anisotropic wireless sensor networks. Wireless Pers. Commun. **98**(4), 3335–3353 (2018). https://doi.org/10.1007/s11277-017-5017-2
34. Sharma, A., Kakkar, A.: Forecasting daily global solar irradiance generation using machine learning. Renew. Sustain. Energy Rev. **82**(August), 2254–2269 (2018). https://doi.org/10.1016/j.rser.2017.08.066
35. Pavlov, Y.L.: Random Forests, pp. 1–122 (2019). https://doi.org/10.1201/9780367816377-11
36. Elghazel, W., Medjaher, K., Zerhouni, N., Bahi, J., Farhat, A., Guyeux, C., Hakem, M.: Random forests for industrial device functioning diagnostics using wireless sensor networks. IEEE Aerosp. Conf. Proc. (2015). https://doi.org/10.1109/AERO.2015.7119275
37. Al-otaibi, H.H.: Associations between sleep quality and different measures of obesity in saudi adults. Glob. J. Health Sci. **9**(1), 1–9 (2017). https://doi.org/10.5539/gjhs.v9n1p1, ISSN 1916-9736, E-ISSN 1916-9744. Published by Canadian Center of Science and Education
38. Praveen Kumar, D., Amgoth, T., Annavarapu, C.S.R.: Machine learning algorithms for wireless sensor networks: a survey. Inf. Fusion **49**(April 2018), 1–25 (2019a). https://doi.org/10.1016/j.inffus.2018.09.013
39. Bailey, T., Jain, A.K.: Note on distance-weighted k-nearest neighbor rules. IEEE Trans. Systems Man Cybern. SMC **8**(4), 311–313 (1978). https://doi.org/10.1109/tsmc.1978.4309958

40. Keller, J.M., Gray, M.R.: A fuzzy K-nearest neighbor algorithm. IEEE Trans. Syst. Man Cybern. SMC **15**(4), 580–585 (1985). https://doi.org/10.1109/TSMC.1985.6313426

41. Sundukov, Y.N.: First record of the ground beetle Trechoblemus postilenatus (Coleoptera, Carabidae) in Primorskii krai. Far Eastern Entomologist **165**(April), 16 (2006). https://doi.org/10.1002/tox

42. Li, Y., Parker, L.E.: Nearest neighbor imputation using spatial-temporal correlations in wireless sensor networks. Inf. Fusion **15**(1), 64–79 (2014). https://doi.org/10.1016/j.inffus.2012.08.007

43. White, H.: Learning in artificial neural networks: a statistical perspective. Neural Comput. **1**(4), 425–464 (1989). https://doi.org/10.1162/neco.1989.1.4.425

44. Chanak, P., Banerjee, I.: Fuzzy rule-based faulty node classification and management scheme for large scale wireless sensor networks. Expert Syst. Appl. **45**, 307–321 (2016). https://doi.org/10.1016/j.eswa.2015.09.040

45. Habib, C., Makhoul, A., Darazi, R., Salim, C.: Self-adaptive data collection and fusion for health monitoring based on body sensor networks. IEEE Trans. Industr. Inf. **12**(6), 2342–2352 (2016). https://doi.org/10.1109/TII.2016.2575800

46. Abu Alsheikh, M., Lin, S., Niyato, D., Tan, H.P.: Rate-distortion balanced data compression for wireless sensor networks. IEEE Sens. J. **16**(12), 5072–5083 (2016). https://doi.org/10.1109/JSEN.2016.2550599

47. Lecun, Y., Bengio, Y., Hinton, G.: Deep learning. Nature **521**(7553), 436–444 (2015). https://doi.org/10.1038/nature14539

48. Lee, Y.M.: Classification of node degree based on deep learning and routing method applied for virtual route assignment. Ad Hoc Netw. **58**, 70–85 (2017). https://doi.org/10.1016/j.adhoc.2016.11.007

49. Lakrami, F., Elkamoun, N., Kamili, M.E.: Advances in ubiquitous networking. Lect. Notes Electr. Eng. **366**, 287–300 (2016). https://doi.org/10.1007/978-981-287-990-5

50. Jain, B., Brar, G., Malhotra, J.: EKMT-k-means clustering algorithmic solution for low energy consumption for wireless sensor networks based on minimum mean distance from base station. Lect. Notes Data Eng. Commun. Technol. **3**, 113–123 (2018). https://doi.org/10.1007/978-981-10-4585-1_10

51. Ray, A., De, D.: Energy efficient clustering protocol based on K-means (EECPK-means)-midpoint algorithm for enhanced network lifetime in wireless sensor network. IET Wireless Sensor Syst. **6**(6), 181–191 (2016). https://doi.org/10.1049/iet-wss.2015.0087

52. Yang, Q., Jang, S.J., Yoo, S.J.: Q-learning-based fuzzy logic for multi-objective routing algorithm in flying ad hoc networks. Wireless Pers. Commun. **113**(1), 115–138 (2020). https://doi.org/10.1007/s11277-020-07181-w

53. Harb, H., Makhoul, A., Couturier, R., Enhanced, A., Harb, H., Makhoul, A.: An enhanced K-means and ANOVA-based clustering wireless sensor networks to cite this version: an enhanced K-means and ANOVA-based underwater wireless sensor networks. IEEE Sens. J. (2019). https://doi.org/10.1109/JSEN.2015.2443380

54. Morell, A., Correa, A., Barceló, M., Vicario, J.L.: Data aggregation and principal component analysis in WSNs. IEEE Trans. Wireless Commun. **15**(6), 3908–3919 (2016). https://doi.org/10.1109/TWC.2016.2531041

55. Anagnostopoulos, C., Hadjiefthymiades, S.: Advanced principal component-based compression schemes for wireless sensor networks. ACM Trans. Sensor Netw. **11**(1), 1–34 (2014). https://doi.org/10.1145/2629330

56. Liu, S., Feng, L., Wu, J., Hou, G., Han, G.: Concept drift detection for data stream learning based on angle optimized global embedding and principal component analysis in sensor networks. Comput. Electr. Eng., 1–10 (2017a). https://doi.org/10.1016/j.compeleceng.2016.09.006

57. Gil, P., Martins, H., Januário, F.: Outliers detection methods in wireless sensor networks. Artif. Intell. Rev. **52**(4), 2411–2436 (2019). https://doi.org/10.1007/s10462-018-9618-2

58. McCall, M.R., Mehta, T., Leathers, C.W., Foster, D.M.: Psyllium husk II: effect on the metabolism of apolipoprotein B in African green monkeys. Am. J. Clin. Nutr. **56**(2), 385–393 (1992). https://doi.org/10.1093/ajcn/56.2.385

59. Almi'ani, K., Viglas, A., Libman, L.: Energy-efficient data gathering with tour length-constrained mobile elements in wireless sensor networks. In: Proceedings—Conference on Local Computer Networks, LCN, pp. 582–589 (2010). https://doi.org/10.1109/LCN.2010.5735777

60. Johnson, S.C.: Hierarchical clustering schemes. Psychometrika **32**(3), 241–254 (1967). https://doi.org/10.1007/BF02289588

61. Neamatollahi, P., Abrishami, S., Naghibzadeh, M., Yaghmaee Moghaddam, M.H., Younis, O.: Hierarchical clustering-task scheduling policy in cluster-based wireless sensor networks. IEEE Trans. Industr. Inf. **14**(5), 1876–1886 (2018). https://doi.org/10.1109/TII.2017.2757606

62. Xu, A., Khokhar, A., Vasilakos, A.V.: Hierarchical data aggregation using compressive sensing (HDACS) in WSNs. ACM Trans. Sensor Netw. **11**(3), 45 (2015). https://doi.org/10.1145/2700264

63. Awan, S.W., Saleem, S.: Hierarchical clustering algorithms for heterogeneous energy harvesting wireless sensor networks. In: Proceedings of the International Symposium on Wireless Communication Systems, 2016 October, pp. 270–274 (2016). https://doi.org/10.1109/ISWCS.2016.7600913

64. Zhu, F., Ma, Z., Zhao, T.: Influence of freeze-thaw damage on the steel corrosion and bond-slip behavior in the reinforced concrete. Adv. Mater. Sci. Eng. (2016). https://doi.org/10.1155/2016/9710678

65. Qin, J., Zhu, Y., Fu, W.: Distributed clustering algorithm in sensor networks via normalized information measures. IEEE Trans. Signal Process. **68**, 3266–3279 (2020). https://doi.org/10.1109/TSP.2020.2995506

66. Nayak, P., Devulapalli, A.: A fuzzy logic-based clustering algorithm for WSN to extend the network lifetime. IEEE Sens. J. **16**(1), 137–144 (2016). https://doi.org/10.1109/JSEN.2015.2472970

67. Guo, P., Cao, J., Liu, X.: Lossless in-network processing in WSNs for domain-specific monitoring applications. IEEE Trans. Industr. Inf. **13**(5), 2130–2139 (2017). https://doi.org/10.1109/TII.2017.2691586

68. Gennarelli, G., Soldovieri, F.: Performance analysis of incoherent RF tomography using wireless sensor networks. IEEE Trans. Geosci. Remote Sens. **54**(5), 2722–2732 (2016). https://doi.org/10.1109/TGRS.2015.2505065

69. Klema, V.C., Laub, A.J.: The singular value decomposition: its computation and some applications. IEEE Trans. Autom. Control **25**(2), 164–176 (1980). https://doi.org/10.1109/TAC.1980.1102314

70. Wold, S., Esbensen, K., Geladi, P.: Decret_Du_7_Mai_1993_Fixant_Les_Modalites_D_Application_De_La_Loi_Relative_Aux_Recensements_Et_Enquetes_Statistiques.Pdf. Chemom. Intell. Lab. Syst. **2**(1–3), 37–52 (1987). https://doi.org/10.1016/0169-7439(87)80084-9

71. Li, X., Ding, S., Li, Y.: Outlier suppression via non-convex robust PCA for efficient localization in wireless sensor networks. IEEE Sens. J. **17**(21), 7053–7063 (2017). https://doi.org/10.1109/JSEN.2017.2754502

72. Oikonomou, P., Botsialas, A., Olziersky, A., Kazas, I., Stratakos, I., Katsikas, S., Dimas, D., Mermikli, K., Sotiropoulos, G., Goustouridis, D., Raptis, I., Sanopoulou, M.: A wireless sensing system for monitoring the workplace environment of an industrial installation. Sens. Actuators B Chemical **224**, 266–274 (2016). https://doi.org/10.1016/j.snb.2015.10.043

73. Stone, J.V.: Independent Component Analysis, pp. 27–66 (2018). https://doi.org/10.7551/mitpress/3717.003.0014

74. Bartlett, M.S., Movellan, J.R., Sejnowski, T.J.: Face recognition by independent component analysis. IEEE Trans. Neural Netw. **13**(6), 1450–1464 (2002). https://doi.org/10.1109/TNN.2002.804287

75. Bravo, C.S., Herrero de Egaña Espinosa de los Monteros, A.: The influences of the downsizing strategy on business structures. Rev. Bus. Manag. **19**(63), 118–132 (2017). https://doi.org/10.7819/rbgn.v19i63.1905

76. Goldberg, X.: Introduction to semi-supervised learning. In: Synthesis Lectures on Artificial Intelligence and Machine Learning, vol. 6 (2009). https://doi.org/10.2200/S00196ED1V01Y200906AIM006

77. Bianchini, M., Maggini, M., Jain, L.C.: Handbook on neural information processing. Intel. Syst. Ref. Libr. **49**, 215–239 (2013). https://doi.org/10.1007/978-3-642-36657-4

78. Kumar, S., Tiwari, S.N., Hegde, R.M.: Sensor node tracking using semi-supervised Hidden Markov models. Ad Hoc Netw. **33**, 55–70 (2015). https://doi.org/10.1016/j.adhoc.2015.04.004

79. Yang, B., Xu, J., Yang, J., Li, M.: Localization algorithm in wireless sensor networks based on semi-supervised manifold learning and its application. Cluster Comput. **13**(4), 435–446 (2010). https://doi.org/10.1007/s10586-009-0118-7

80. Yoo, J., Jin Kim, H.: Target localization in wireless sensor networks using online semi-supervised support vector regression. Sensors (Switzerland) **15**(6), 12539–12559 (2015). https://doi.org/10.3390/s150612539

81. Zhao, M., Chow, T.W.S.: Wireless sensor network fault detection via semi-supervised local kernel density estimation. In: Proceedings of the IEEE International Conference on Industrial Technology, 2015 June, pp. 1495–1500 (2015). https://doi.org/10.1109/ICIT.2015.7125308

82. Baird, L.: Residual algorithms: reinforcement learning with function Approximation. In: Machine Learning Proceedings 1995. Morgan Kaufmann Publishers, Inc. (1995). https://doi.org/10.1016/b978-1-55860-377-6.50013-x

83. Guo, W., Yan, C., Lu, T.: Optimizing the lifetime of wireless sensor networks via reinforcement-learning-based routing. Int. J. Distr. Sensor Netw. **15**(2) (2019). https://doi.org/10.1177/1550147719833541

84. Lu, Y., He, R., Chen, X., Lin, B., Yu, C.: Energy-efficient depth-based opportunistic routing with q-learning for underwater wireless sensor networks. Sensors (Switzerland), **20**(4) (2020). https://doi.org/10.3390/s20041025

85. Ping, W., Ting, W.: Adaptive routing for sensor networks using reinforcement learning. In: Proceedings—Sixth IEEE International Conference on Computer and Information Technology, CIT 2006, p. 219 (2006). https://doi.org/10.1109/CIT.2006.34

86. Dong, S., Agrawal, P., Sivalingam, K.: Reinforcement Learning Geo Routing Protocol WSNs, pp. 652–656. IEEE (2007). https://doi.org/10.1109/GLOCOM.2007.127

87. Förster, A., Murphys, A.L.: FROMS: feedback routing for optimizing multiple sinks in WSN with reinforcement learning. In: Proceedings of the 2007 International Conference on Intelligent Sensors, Sensor Networks and Information Processing, ISSNIP, vol. 5005, pp. 371–376 (2007). https://doi.org/10.1109/ISSNIP.2007.4496872

88. Kulkarni, S.R., Lugosi, G., Venkatesh, S.S.: Learning pattern classification—a survey. IEEE Trans. Inf. Theory **44**(6), 2178–2206 (1998). https://doi.org/10.1109/18.720536

89. Khan, Z.A., Samad, A.: A study of machine learning in wireless sensor network. Int. J. Comput. Netw. Appl. **4**(4), 105–112 (2017). https://doi.org/10.22247/ijcna/2017/49122

90. Youssry, N., Khattab, A.: Ameliorating IoT and WSNs via machine learning. In: Proceedings of the International Conference on Microelectronics, ICM, 2019-Decem, pp. 342–345 (2019). https://doi.org/10.1109/ICM48031.2019.9021276

91. Han, G., Xu, H., Duong, T.Q., Jiang, J., Hara, T.: Localization algorithms of wireless sensor networks: a survey. Telecommun. Syst. (2011). https://doi.org/10.1007/s11235-011-9564-7

92. Shi, K., Ma, Z., Zhang, R., Hu, W., Chen, H.: Support Vector Regression Based Indoor Location in IEEE 802.11 Environments (2015)

93. Bhatti, G.: Machine learning based localization in large-scale wireless sensor networks. Sensors (Switzerland) **18**(12) (2018). https://doi.org/10.3390/s18124179

94. Baccar, N., Bouallegue, R.: Interval type 2 fuzzy localization for wireless sensor networks. Eurasip J. Adv. Signal Process. **2016**(1) (2016). https://doi.org/10.1186/s13634-016-0340-4

95. Barbancho, J., León, C., Molina, F.J., Barbancho, A.: A new QoS routing algorithm based on self-organizing maps for wireless sensor networks. Telecommun. Syst. **36**(1–3), 73–83 (2007). https://doi.org/10.1007/s11235-007-9061-1

96. Kim, M.H., Park, M.G.: Bayesian statistical modeling of system energy saving effectiveness for MAC protocols of wireless sensor networks. Stud. Comput. Intell. **209**, 233–245 (2009). https://doi.org/10.1007/978-3-642-01203-7_20

97. Kulkarni, R.V., Venayagamoorthy, G.K.: Neural network based secure media access control protocol for wireless sensor networks. In: Proceedings of the International Joint Conference on Neural Networks, pp. 1680–1687 (2009). https://doi.org/10.1109/IJCNN.2009.5179075

98. Chu, Y., Mitchell, P.D., Grace, D.: ALOHA and Q-learning based medium access control for wireless sensor networks. In: Proceedings of the International Symposium on Wireless Communication Systems, pp. 511–515 (2012). https://doi.org/10.1109/ISWCS.2012.6328420

99. Sha, M., Dor, R., Hackmann, G., Lu, C., Kim, T.S., Park, T.: Self-adapting MAC layer for wireless sensor networks. Proc. Real Time Syst. Symp. 192–201 (2013). https://doi.org/10.1109/RTSS.2013.27

100. Praveen Kumar, D., Amgoth, T., Annavarapu, C.S.R.: Machine learning algorithms for wireless sensor networks: a survey. Inf. Fusion **49**(April 2018), 1–25 (2019b). https://doi.org/10.1016/j.inffus.2018.09.013

101. Ahmed, G., Khan, N.M., Khalid, Z., Ramer, R.: Cluster head selection using decision trees for wireless sensor networks. In: ISSNIP 2008—Proceedings of the 2008 International Conference on Intelligent Sensors, Sensor Networks and Information Processing, pp. 173–178 (2008). https://doi.org/10.1109/ISSNIP.2008.4761982

102. Betta, G., Casinelli, D., Ferrigno, L.: Some notes on the performance of regression-based time synchronization algorithms in low cost WSNs. Springer International Publishing, Cham (2015)

103. Kosunalp, S.: A New Energy Prediction Algorithm for Energy—Harvesting Wireless Sensor Networks with Q—Learning, vol. 3536(c) (2016). https://doi.org/10.1109/ACCESS.2016.2606541

104. Shaikh, F.K., Zeadally, S.: Energy harvesting in wireless sensor networks: a comprehensive review. Renew. Sustain. Energy Rev. **55**, 1041–1054 (2016). https://doi.org/10.1016/j.rser.2015.11.010

105. Moustapha, A.I., Selmic, R.R.: Wireless sensor network modeling using modified recurrent neural networks: application to fault detection. IEEE Trans. Instr. Meas, 15–17 (2007). https://doi.org/10.1109/TIM.2007.913803

106. Balouchestani, M., Raahemifar, K., Krishnan, S.: Compressed sensing in wireless sensor networks: survey. In: Compressed Sensing in Wireless Sensor Networks: Survey, January 2011 (2015)

107. Kimura, N., Latifi, S.: A survey on data compression in wireless sensor networks. Int. Conf. Inf. Technol. Coding Comput. ITCC **2**, 8–13 (2005). https://doi.org/10.1109/itcc.2005.43

108. Yu, D., Guo, Y., Li, N., Wang, M.: SA-M-SBL: an algorithm for CSI-based device-free localization with faulty prior information. IEEE Access **7**, 61831–61839 (2019). https://doi.org/10.1109/ACCESS.2019.2916194

109. Almuzaini, K.K., Gulliver, T.A.: Range-Based Localization in Wireless Networks Using Decision Trees, pp. 131–135 (2010)

110. Xu, Y., Ding, O., Qu, R., Li, K.: Hybrid multi-objective evolutionary algorithms based on decomposition for wireless sensor network coverage optimization. Appl. Soft Comput. J. **68**, 268–282 (2018). https://doi.org/10.1016/j.asoc.2018.03.053

111. Kim, W., Member, S., Stankovi, M.S., Johansson, K.H., Kim, H.J.: Over wireless sensor networks. IEEE Trans. Cybern. **45**(11), 1–13 (2015)

112. Chang, X., Huang, J., Liu, S., Xing, G., Zhang, H., Wang, J.: Accuracy-Aware Interference Modeling and Measurement in Wireless Sensor Networks, vol. 1233(c), pp. 1–14 (2015). https://doi.org/10.1109/TMC.2015.2416182
113. Chen, H., Li, X., Zhao, F.: A Reinforcement Learning-Based Sleep Scheduling Algorithm for Desired Area Coverage in Solar-Powered Wireless Sensor Networks (2016). https://doi.org/10.1109/JSEN.2016.2517084
114. Arroyo-Valles, R., Marqués, A.G., Vinagre-Díaz, J.J., Cid-Sueiro, J.: A Bayesian decision model for intelligent routing in sensor networks. In: 3rd International Symposium on Wireless Communication Systems 2006, ISWCS'06, pp. 103–107 (2006). https://doi.org/10.1109/ISWCS.2006.4362268
115. Turčaník, M.: Advances in military technology. Neural Netw. Approach Routing Sensor Netw. 8(2), 71–82 (Advances in Millitary Technology) (2013)
116. Rovcanin, M., Poorter, E.D., Moerman, I., Demeester, P.: Ad hoc networks a reinforcement learning based solution for cognitive network cooperation between co-located, heterogeneous wireless sensor networks. Ad Hoc Netw. 17, 98–113 (2014). https://doi.org/10.1016/j.adhoc.2014.01.009
117. Phung, K., Lemmens, B., Goossens, M., Nowe, A., Tran, L., Steenhaut, K.: Ad Hoc Networks schedule-based multi-channel communication in wireless sensor networks: a complete design and performance evaluation. Ad Hoc Netw. 26, 88–102 (2015). https://doi.org/10.1016/j.adhoc.2014.11.008
118. Kosunalp, S., Chu, Y., Mitchell, P.D., Grace, D., Clarke, T.: Engineering applications of artificial intelligence use of Q-learning approaches for practical medium access control in wireless sensor networks. Eng. Appl. Artif. Intell. 55, 146–154 (2016). https://doi.org/10.1016/j.engappai.2016.06.012
119. Mustapha, I., Ali, B.M., Sali, A., Rasid, M.F.A.: An energy efficient reinforcement learning based cooperative channel sensing for cognitive radio sensor networks. Pervasive Mob. Comput. (2016). https://doi.org/10.1016/j.pmcj.2016.07.007
120. Li, Y., Parker, L.E.: Nearest neighbor imputation using spatial—temporal correlations in wireless sensor networks. Inf. Fusion (2012). https://doi.org/10.1016/j.inffus.2012.08.007
121. Bertrand, A., Moonen, M.: Distributed adaptive estimation of covariance matrix eigenvectors in wireless sensor networks with application to distributed PCA $. Sig. Process. 104, 120–135 (2014). https://doi.org/10.1016/j.sigpro.2014.03.037
122. Chidean, M.I., Morgado, E., Arco, E., Ramiro-Bargue, J., Caama, A.J.: Scalable Data-Coupled Clustering for Large Scale WSN, vol. X, issue X, pp. 1–13 (2015). https://doi.org/10.1109/TWC.2015.2424693
123. Liu, S., Feng, L., Wu, J., Hou, G., Han, G.: Concept drift detection for data stream learning based on angle optimized global embedding and principal component analysis in sensor networks. Comput. Electr. Eng. 58, 327–336 (2017b). https://doi.org/10.1016/j.compeleceng.2016.09.006
124. Edwards-murphy, F., Magno, M., Whelan, P.M., Halloran, J.O., Popovici, E.M.: b+WSN: smart beehive with preliminary decision tree analysis for agriculture and honey bee health monitoring q. Comput. Electron. Agric. 124, 211–219 (2016). https://doi.org/10.1016/j.compag.2016.04.008
125. Atoui, I., Makhoul, A., Tawbe, S.: Tree-Based Data Aggregation Approach in Periodic Sensor Networks Using Correlation Matrix and Polynomial Regression (2016). https://doi.org/10.1109/CSE-EUC-DCABES.2016.267
126. Gispan, L., Leshem, A., Be, Y.: Decentralized estimation of regression coefficients in sensor. Digit. Signal Proc. 68, 16–23 (2017). https://doi.org/10.1016/j.dsp.2017.05.005
127. Das, S.K.: An Adaptive Bayesian System for Context-Aware Data Fusion in Smart Environments (2016). https://doi.org/10.1109/TMC.2016.2599158

128. Hwang, S., Member, S., Ran, R., Yang, J.: Multivariated Bayesian Compressive Sensing in Wireless Sensor Networks, pp. 1–10 (2015). https://doi.org/10.1109/JSEN.2015.2508670

129. Wang, C., Bertino, E.: Sensor network provenance compression using dynamic Bayesian networks. ACM Trans. Sensor Netw. 13(1) (2017). https://doi.org/10.1145/2997653

130. Alsheikh, M.A., Member, S., Lin, S.: Rate-Distortion Balanced Data Compression for Wireless Sensor Networks, pp. 1–12 (2016). https://doi.org/10.1109/JSEN.2016.2550599

131. Rezaee, A.A.: A fuzzy congestion control protocol based on active queue management in wireless sensor networks with medical applications. Wireless Pers. Commun. (2017). https://doi.org/10.1007/s11277-017-4896-6

132. Braca, P., Willett, P., Lepage, K., Marano, S., Matta, V.: Bayesian Tracking in Underwater Wireless Sensor Networks With Port-Starboard Ambiguity, vol. 62, issue 7, pp. 1864–1878 (2014)

133. Das, S.N., Misra, S., Member, S., Member, B.E.W.: Temporal-Correlation Aware Dynamic Self-Management of Wireless Sensor Networks, vol. 3203(c), pp. 1–13 (2016). https://doi.org/10.1109/TII.2016.2594758

134. Wei, Z., Zhang, Y., Xu, X., Shi, L., Feng, L.: A task scheduling algorithm based on Q-learning and shared value function for WSNs. Comput. Netw. 126, 141–149 (2017). https://doi.org/10.1016/j.comnet.2017.06.005

135. Avci, B., Trajcevski, G., Tamassia, R., Scheuermann, P., Zhou, F.: Efficient Detection of Motion-Trend Predicates in Wireless Sensor Networks. vol. 101, pp. 26–43 (2017). https://doi.org/10.1016/j.comcom.2016.08.012

136. Ye, D., Zhang, M.: A self-adaptive sleep/wake-up scheduling approach for wireless sensor networks. IEEE Trans. Cybern. 979–992 (2018). https://doi.org/10.1109/TCYB.2017.2669996

137. Wu, M., Feng, Q., Wen, X., Deo, R.C., Yin, Z.: Uncorrected Proof Oasis Region Uncorrected Proof, pp. 1–18 (2020). https://doi.org/10.2166/nh.2020.012

138. Zhang, R., Pan, J., Member, S., Xie, D., Member, S., Wang, F.: NDCMC: A Hybrid Data Collection Approach for Large-Scale WSNs Using Mobile Element and Hierarchical Clustering (2015). https://doi.org/10.1109/JIOT.2015.2490162

139. Kim, S., Kim, D.: Efficient data-forwarding method in delay-tolerant P2P networking for IoT services. Convergence P2P Cloud Computing. Springer (2017)

140. Banimelhem, O., Abu-hantash, A.: Fuzzy logic-based clustering approach with mobile sink for WSNs. In: 13th International Computer Engineering Conference (ICENCO), pp. 36–40 (2017). https://doi.org/10.1109/ICENCO.2017.8289759

141. Hsu, R.O.Y.C., Liu, C., Wang, H.: A reinforcement learning-based ToD provisioning dynamic power management for sustainable operation of energy harvesting wireless sensor node. IEEE Trans. Emerg. Topics Comput. 2(2), 181–191 (2014). https://doi.org/10.1109/TETC.2014.2316518

142. Aoudia, F.A., Gautier, M., Berder, O.: RLMan: An Energy Manager Based on Reinforcement Learning for Energy Harvesting Wireless Sensor Networks, vol. 2400(c), pp. 1–11 (2018). https://doi.org/10.1109/TGCN.2018.2801725

143. Wazid, M., Das, A.K.: An efficient hybrid anomaly detection scheme using k-means clustering for wireless sensor networks. Wireless Pers. Commun. 90(4), 1971–2000 (2016). https://doi.org/10.1007/s11277-016-3433-3

144. Meng, W., Li, W., Xiang, Y., Choo, K.R.: Author's accepted manuscript a Bayesian inference-based detection mechanism to defend medical smartphone networks against insider attacks reference: a Bayesian inference-based detection mechanism to defend medical. J. Netw. Comput. Appl. (2016). https://doi.org/10.1016/j.jnca.2016.11.012

145. Titouna, C., Aliouat, M., Gueroui, M.: FDS: fault detection scheme for wireless sensor. Wireless Pers. Commun. (2015). https://doi.org/10.1007/s11277-015-2944-7

146. Cheng, Y., Liu, Q., Wang, J., Wan, S.: Distributed Fault Detection for Wireless Sensor Networks Based on Support Vector Regression (2018)

147. Yuan, X., Member, S.: WNN-LQE: wavelet-neural-network-based link quality estimation for smart grid WSNs. IEEE Access **5**, 12788–12797 (2017). https://doi.org/10.1109/ACCESS.2017.2723360
148. Chandrakala, A.P.R.S.: MRL-SCSO: multi-agent reinforcement learning-based self-configuration and self-optimization protocol for unattended wireless sensor networks. Wireless Pers. Commun. (2016). https://doi.org/10.1007/s11277-016-3729-3
149. Collotta, M., Pau, G., Bobovich, A.V.: A Fuzzy Data Fusion Solution to Enhance the QoS and the Energy Consumption in Wireless Sensor Networks. Wireless Communications and Mobile Computing (2017). https://doi.org/10.1155/2017/3418284

Data Analysis for Smart Sensor Networks

Energy Efficient Smart Lighting System for Rooms

Madhavi Vaidya, Pritish Chatterjee, and Kamlakar Bhopatkar

Abstract In the twenty-first century, the demand of power has gone up and this requires more generation of power. This leads to faster consumption of raw materials and results in more pollution. The need for the present time is for a smart lighting system that can adjust itself with respect to the natural light in order to save energy. In recent years, there has been a growing concern about energy consumption within residential buildings. Currently, two main strategies are available for reducing energy consumption by lighting, according to Martirano (Smart lighting control to save energy, 132–138, 2011) [1] either increasing efficiency or effectiveness. The efficiency improves by applying more efficient light sources while effectiveness means the implementation of automated control systems including, for instance, daylight harvesting or occupancy sensors. For home environments, the energy-efficient light sources currently available on the market are Compact Fluorescent Lights (CFLs), Light Emitting Diodes (LEDs), and Smart LEDs. In this chapter, a smart lighting system would be exemplified that can control the room-light efficiently by using sensors to dim and brighten whenever it is required. This system is based on the concept of IoT (Internet of Things). When the system will be operating under automatic mode, the system will maintain the intensity set by the user with respect to the natural light by either increasing or decreasing the artificial lighting. In case of automatic usage of the lighting system, the AI system will be able to predict the suitable lighting intensity based on the previous records of intensity manually set by the user. It will take in features like time, weather and natural light intensity recorded by the LDR. This saves the user's time as he/she may simply need to choose between the suggested intensity by the system or choose to set the intensity himself/herself. To make the system truly intelligent, a smart camera system and IR sensor would be installed that would command the lighting system to adjust the lighting intensity according to the user's selected choice when entering the room. The two systems have been studied in this chapter and the systems are, Self-Adjusting Lighting System and the other is Facial Recognition based Lighting Management System. The first system brings automation to homes and saves time and energy whereas the later one uses Artificial

M. Vaidya (✉) · P. Chatterjee · K. Bhopatkar
Vivekanand Education Society's College of Arts, Science & Commerce, Mumbai, India

© The Author(s), under exclusive license to Springer Nature Switzerland AG 2022
U. Singh et al. (eds.), *Smart Sensor Networks*, Studies in Big Data 92,
https://doi.org/10.1007/978-3-030-77214-7_5

Intelligence to bring about energy efficiency and make homes smarter. The aim of this chapter is to automate the lights of the room to increase the productivity and accuracy of the system in a cost-effective manner and also permits wireless accessibility and control over the system.

Keywords Arduino. · LDR · IoT · Smart lighting systems · Lighting

1 Introduction

Energy is major input sector for economic development of any country. Smart Home is the term commonly used to define a residence that participates technology and services through home networking to enhance power effectiveness and improve the quality of living [2]. Smart Home is technology to make a house to become bright and automated. Smart lighting systems mainly involve digital sensors, actuators drivers and communications interfaces [3]. Usually, that technology has automation systems for lighting, temperature control, security and many other functions [4]. According to Robles and Kim [5], Smart Home is the term for defining residence using the control system to participate home automation system. The system allows integrating electronic devices controller with only a few buttons that are connected with the simple telecommunications system. Smart Home includes communications, entertainment, security, convenience, and information systems [6]. There are several terms that are corresponding with the Smart Home such as: Home Automation, Intelligent Building or Home Networking. Home automation is the automatic control of electronic devices utilized at home. These devices are connected to the Internet, which allows them to be controlled remotely. With home automation, devices can trigger one another and there is no need to control them manually via an app.

The purpose of home automation is to make homes simpler, organized, and more available which can have smart lighting system. Home automations is not one technology, it's the integration of multiple technologies into one system. Smart lighting uses intelligent, connected devices to make buildings more sustainable, responsive, secure, responsive and healthy for the people in them. It uses networked lighting sensors to control all lighting throughout a building or campus. Smart home ensures and allows the users to save electricity and reduce your power and water bills.

The global lighting market has been undergoing a great transformation driven by the massively growing adoption of light emitting diode (LED) technology. In many conventional applications, LEDs are used and those are expensive compared with other light sources. One of the drawbacks of a conventional lighting system is that it lacks the flexibility for any relocation of light sources. Even it requires a great effort to rewire the entire system once it gets big, may be a building having many more floors or heighted etc. In a conventional lighting system, manual ON/OFF of a light source is carried out, while, instead in a smart one, various present lighting modes are preloaded into the lighting system, either wired or wireless, to meet the user's specific needs. Besides, conventionally, a heavily loaded lighting system requires a high-capacity switch, and requires a large volume of cables to drive a distant load. Actually, a load is directly powered by an output driver, meaning that there is no need to increase the power capacity of a switch when the system is heavily loaded, and it merely requires a long signal line to drive a distant load [7].

These days, the instant energy use in lighting in such a high-rise building must be monitored in real time for energy saving purposes. A smart lighting system refers to an MCU-based system integrating automation, electronics, computer, network communication, and many more for energy efficiency improvement. Cost of operating automatic solar street lights is far less when compared to the conventional street lights. There are lower chances of the automatic street light system over-heating and risk of accidents is also minimized.

Besides all these points which have been explained on top, a smart lighting system can be made dimmable and controllable which is manageable by timer RC. It is a combination of home automation using LDR and IR sensor is a further step in home automation.

The system also uses IR sensors to control or to operate household appliances like TV, fans, music systems, tube light, radio within the range of 10 m. IR sensor is a type of sensor, that emits the light inorder to sense some object of the surroundings. An IR sensor can measure the heat of an object as well as detects the motion. Various hardware parts are required for RC Automation are IR transmitter, IR receiver, transistors, IC, LED light, LDR sensor, batteries, bread board and connecting wires. The system works as soon as the IR rays by the IR transmitter are sensed by the IR receiver. The response of the IR receiver is specified that will turn on or will turn off the device accordingly. The IR transmitter is nothing but Remote controller of TV or of media player, The IR receiver module of the circuit can be embedded with home appliance which the person wants to control for making the home automation suitable, convenient and gives an appropriate performance.

The IR sensors are very effective and give the good coverage. It will work as soon as it senses the light and turn off the light. LDR (light dependent resistor) is a light dependent sensor, it detects the light as the retina of a human eye does mean.

A proximity sensor is a sensor able to detect the presence of nearby objects without any physical contact. We can say that a proximity sensor is a non-contact sensor that detects the presence of an object or target when it happens that the target enters the sensor's field. Proximity sensors are also used in machine vibration monitoring to measure the variation in distance between a shaft and its support bearing.

2 Techniques and Tools for Smart Lighting System Sensors

Various techniques are used for smart lighting system viz. IoT, LDR, Home Intelligence and few more like occupancy sensing based system.

2.1 IoT

A smart city in an environment and infrastructure which is highly dependent upon Internet for communication and services. Thus, IoT is a key factor for building smart cities. Microcontroller (MCU) that processes data and runs software stacks interfaced to a wireless device for connectivity.

The smart home system is able to sense its environment and accordingly send alerts to the user on registered device or account. The information related to environmental data which may include temperature, humidity, light intensity etc. alert may be sent to user on regular basis at predefined time. Alert may be sent over email, as a text message, through tweets or through any other social media.

The most important purpose of smart home is monitoring and keeping track of every activity in a smart home which is the primary need on basis of which any further action can be taken or decision can be made. Tracking and taking care of room temperature and sending alert to user to switch on air-conditioner if temperature is above certain degree, can be an example of the said monitoring system.

2.2 LDR

LDRs (light-dependent resistors) are used to detect light levels used in automatic security lights. LDR is a sensor that has a changing resistance that changes with the light intensity that falls upon it. LDR has application in light sensing circuits. As the resistance decreases as the light intensity increases: in the dark and at low light levels, the resistance of an LDR is high and little current can flow through it.

Light dependent resistors or photo resistors are mostly used in circuits where it is necessary to detect the presence or the level of light. LDR is made up of semiconductor materials with high resistance. The working principle of LDR is based on the photo conductivity that is an optical phenomenon [8]. LDR's resistance is described when light falls on them. It increases when the room is dark and LDR's residency will be as high as Ω. It decreases when there is daylight or sunlight. So, if a constant voltage is applied on the LDR and light intensity is increased then the current will start increasing as well.

2.3 Home Intelligence (HI)

It is the most significant function of smart home and refers to intelligent behaviour of the smart-home environment. This function is related to automatically making decision on occurrence of various events. HI depends upon the Artificial Intelligence (AI) mechanism built in the smart home environment. HI does is also very important for security point of view in a home [9].

2.4 Occupancy-Sensing Based Systems

Occupancy sensors work by detecting the motion and changes in the environment; various sensing technologies carry it out in a different way. It is a popular energy-saving method due to its effectiveness and it is very easy to implement. Occupancy sensing systems allows for dimming or turning off lights automatically after a given space has been vacant for a certain user-specified period of time. Once the movement is detected in the specific room, lights get turned on or dim up as required. In fact, in North American and European building codes the occupancy detection technologies have been heavily endorsed [10, 11]. The integration of simple occupancy sensors is dependent on occupants' usage patterns and it has been found that 3–60% energy savings can be made [12].

Generally, it is noticed that the disadvantage of this occupancy detection technologies are generally dependent on single-point sensing technology, where the data is collected by a sensor detection system which is in the specific room and it has been found that it is neither shared with other room detection systems nor not utilized for the later analysis if required [13]. This might possibly generate significant uncertainty in the sensor feedback data.

It has been stated in the literature that, this occupancy-based sensing system generally results in increased energy wastage. There are chances that the longer time delay happens and that results in less disruption of occupant activity but a higher power consumption and vice versa. Another challenge faced by this type of sensing systems is, the limited detection field-of-view. Many a times it happens that the lights are switched off in the areas or spaces that are occupied as the user is situated outside the sensor coverage area and sometimes that provokes the users to disable the sensors that are configured in the specific are [13].

2.5 Sensor Technology

This is [14] also used for the same objective of conserving energy. Vacancy Sensor can also be a replacement for standard wall switches. By using the passive infrared technology (PIR), these sensors combine occupancy detection and voltage switching in a single package. These units automatically sense and turn off lights when a room is detected as vacant for 5–10 min. The sensors those are mounted on ceilings also use passive infrared technology which detects vacancy and turn OFF lights automatically. They have a 180 degree and a 360-degree field of view and can cover up to 1000 square feet of area. However, sensor technology has few drawbacks as, sensors are more expensive and are likely to break. Moreover, sensors can sense objects/people to a limited range i.e. one sensor might not cover a full room and also it requires lot of additional wiring in case of wired sensors.

2.6 Arduino Board

Arduino is a mechanism and an open-source physical computing platform based on a simple microcontroller board for making computers sense and control more of the physical world than your desktop computer. It is a development environment for writing software for the board. Arduino can be used to develop interactive objects, taking inputs from a variety of switches or sensors, and controlling a variety of lights, motors, and other physical outputs. Arduino projects can be standalone, or communicate with software running on your computer. There are many other microcontrollers and microcontroller platforms available for physical computing. Arduino also manages the process of working with microcontrollers, but it offers advantage to teachers, students, and interested amateurs over other systems [15].

3 Commercial Smart Lighting Systems

The vital controllability of LEDs, along with their easily marketable high energy effectiveness are among the factors that carried out major lighting companies to invest in the development of smart lighting systems. First reason behind opening the doors for the LEDs is the brightness of LEDs and secondly LED lights consume less electricity than standard incandescent.

We can see that commercial smart lighting systems making their mark on the global lighting market.

In the following paragraph, there is a reference that gives an overview of the commercial smart lighting systems. If we take an example of the Hue system that consists of wireless RGB LED bulbs and a wireless control bridge of ZigBee. The major selling point of the innovative Hue bulb was its newness; it introduced the concept of lighting control beyond that of a simple on/off switch to consumers around the world.

By looking at the success of the Hue, few other companies like OSRAM and GE also began to invest in smart lighting systems. Today, there are a vast range of various smart lighting solutions are present in the market viz. LIFX, Belkin WeMo, GE Link, OSRAM Lightify, etc. and few other companies as mentioned in Table 1. The working of all these lighting systems that are smart in nature, is quite similar to the original Philips Hue; only there are slight changes in the method of implementation, depending on the number of LED channels and the connectivity methods differ. One more important fact about the commercial smart lighting products is they include a free cross platform mobile application that functions as a user interface to allow various users to adjust the output of light generated and control various parameters. Following table shows a summary and even a comparative approach for some of the popular commercial smart lighting products. One notable and important similarity of these products is the strong emphasis on aesthetics and mobile control [16].

Table 1 Commercial lighting system [16]

Product	Connectivity	Availability of mobile app	Hub
OSRAM lightify	ZigBee + WiFi	✓	✓
Elgato Avea	Bluetooth	✓	X
LIFX	Mesh type of WiFi network	✓	✓
GE link smart LED	ZigBee + WiFi	✓	✓
Belkin WeMo	ZigBee + WiFi	✓	✓
LightWave RF	WiFi + LightWave RF	✓	X
Philips Hue	ZigBee + WiFi	✓	✓

4 Overview of Lighting System

In order to know more about conventional lighting system which is nothing but, the more information on solid state lighting, how white light can be generated for illumination and/or communication purposes. Humans are adapted to working in healthy environments that would give us sun daylight spectrum. For this reason, we always seek to illuminate a space with white light that can be received from solar spectrum. There is a way to receive an artificial white light with LEDs and both techniques are of interest in human centric applications (HCA) and optical communications where perception of colour and light spectrum are very important parameters. The latest LED technology opens up wide areas for new applications, new technical potentials and reduced costs. The first technique to obtain white light for illumination purposes, or optical communications, that employs a combination of red, green and blue (RGB) LEDs [17, 18].

The second method is to obtain a white light, it employs a phosphorous coating in contact with LED encapsulation resin [19].

However, despite there are few differences in light quality, acceptance by user and the cost part, compared to CFLs, LEDs are much more energy efficient. Smart lighting system (SLS), requiring Smart LEDs, supposed to be even more efficient than conventional lighting systems (using CFL or LED) by using protocols or present modes while connected and adjusted via wireless networks for user needs and requirements.

5 Review of Literature Methods of Smart Lighting Systems and Various Case Studies

At the University of California, Kim et al. [20] present SPOTLIGHT, one of the system which is a prototype that can check and monitor energy consumption by individuals using a proximity sensor. A preliminary idea behind a proximity sensor is that an occupant carries an active RFID tag, which is used for detecting proximity between a user and each appliance. This proximity information is then used for energy loss, reporting the energy consumption profile for useful power or wasted power of each user who has used an appliance (e.g. TV, lamps or home appliances etc.)

We studied one more system mentioned in [21], the authors describe about automated lighting system with visitors' counter. This System has an automatic working of the system. It does not need any manual operation for switching ON/OFF when a person enters or exits from a room. The IR sensors with the IR transmitter and receiver are placed at the entrance of the room doors in such a way that the sensor would sense a person at the entry point or while exiting the room. This has been implemented using a laser. A Microcontroller that is used in controlling the lights and fans in a room and keeps track of number of visitors entering or leaving the room.

It has been implemented in this system in such a way that when a person enters into the room then the counter is incremented by one and the lights in the room will be switched ON and when a person leaves the room then the counter is decremented by one.

The lights will only be switched OFF until all the persons who were present in the room go out and the room is unoccupied. However, the limitation of this system is that the room doors do not get opened wide enough so that two or more people should not be allowed to enter at the same time [21].

In the IR Based Home Appliances Control System [22], the device is able to control different home appliances within a particular range. The circuit is connected to any home appliances (lamp, fan, TV and various home appliances etc.) to make the appliance switch on/off and regulate the speed of fans using a remote control. It can be activated from up to 10 m. It has been mentioned that instead of going near the appliance to turn it on/off we just need to press a button. It has been exemplified as this system is very easy to install and can be assembled on the back side of a switchboard. IR-Based Home Appliances Control System is tried to develop which is used to switch the state of an appliance using a remote. The target of this system is to make the life of people easy and save electricity [22].

Another study has been carried out by Centre for Energy Studies of Indian Institute of Technology have designed a smart occupancy sensor which understands and senses the variation in activity level of the occupants with respect to time of the day. With this information, the system can change the time delay (TD) with the time of the day. Thus, more energy can be saved as compared to non-adapting fixed time delay sensors [23].

Zhen et al. in [24] have implemented a system with multiple active RFID readers, and developed a localization algorithm based on support vector machine (SVM), by giving insight on lighting control for energy saving.

This study in [25]. Comparison of simulated energy consumption by smart and conventional lighting systems in a residential setting. In [26] is intended to compare simulated Electric Lighting Energy Consumption (ELEC) for two different types of lighting systems like Compact Fluorescent Lights and Smart lighting system (CFL, LED and SLS). The findings show that SLS consumes the minimum ELEC and has a higher energy saving potential compared to CFL and LED. This can be explained by the placement of sensors, Smart LEDs, and pre-defined settings based on occupancy pattern. For example, the General Lighting (GL) level in SLS is automatically lowered in rooms while in the other lighting system (using CFL or LED) keeps the lights on. Furthermore, by using SLS and occupancy sensors, lights can automatically be turned off when the rooms are not occupied.

It is clear that as per the various literature that has been studied there is still a lot of scope for development and expansion in front of the various smart lighting products to become as an option that would be accepted by the various users who can use it normally. Such type of lighting systems are needed to be considerably developed to have various useful functionalities such as improved energy competency, better quality of lighting, finer use of every day's functionality, all such factors are needed to be present inside the same system.

6 Comparative Approach of the Implementation Mechanisms

Until now we understood and looked at an exemplification of working of Smart Homes. Smart home appliance is an interface between the remote control with its mobile or remote control and a home reliever.

For each device, in order to accomplish the interface design of smart homes, the micro controller and Arduino are used for controlling some application in the home manually by using a remote control and automatically through different sensors. Arduino UNO is a microcontroller board it has 14 digital input/output pins (of which 6 can be used as PWM outputs), 6 analog inputs, a 16 MHz ceramic resonator, a USB connection, a power jack, an ICSP header, and a reset button. It contains everything needed to support the microcontroller; simply connect it to a computer with a USB cable or power it with an AC-to-DC adapter or battery to get started.

6.1 Smart Home System Based on DTMF Technology

Dual Tone Multi Frequency. (DTMF) tones generated from cell phones or mobile phones' keypad control home devices and appliances remotely [27].

The system used specific signal from mobile phones digits to perform specific task. The signal generated from a DTMF keypad consists of two frequencies: column frequency is high in nature whereas row frequency is low. The unique tone generated by the digits of a DTMF keypad enables the system to perform previously specified task automatically. The system requires a mobile phone and a DTMF transmitter to send signal.

6.2 Smart Home System Based on GSM

Home automation system where GSM is used, it requires a mobile phone, GSM module, microcontroller board and control circuit to control appliances of home [27, 28] Commands are sent via SMS to GSM module which receives the message and sends it to the microcontroller board to implement the various commands. With the help of relay module in it, the microcontroller board turns on/off the specific appliances which are used in smart home. In this way, remote switching of home appliances can be done using this type of GSM module. Smart Home presented in [27] has achieved more than 98% accuracy and the entire processing of sending and receiving commands was done within 2 s. Smart Home based on GSM technology presented in [28] comprised of an LCD which displayed important messages to the user.

6.3 Smart Home System Based on Voice Recognition

In case of voice recognition-based automation, Zigbee based smart home systems are most popular [29]. Zigbee based systems can be divided into 3 major modules-microphone module, Zigbee coordinator (central controller) and Terminals (appliance controller). The system requires a smart device like smart phone or computer which is connected to Zigbee coordinator module. The Zigbee coordinator module is branched into different terminals. The terminals can complete different specified task such as monitoring and control of temperature, gas, humidity, the switching of various appliances and many other functionalities. To perform these tasks, different sort of sensors and appliance control circuit are needed to be installed.

6.4 Smart Home System Based on Wi-Fi and Internet

Home Automation System based on Internet is speedily gaining importance due to its wide range of functions [30]. IoT Based Smart Home System developed in [31] incorporates smart control of lighting, various intelligent appliances control, intrusion alarm system and gas/smoke detection. Home appliance monitoring and controlling system based on IoT developed in [32].

Authors [33] used a software application on smart phone, does controlling of various devices. This system is divided into 3 major parts: remote environment, home gateway and home environment. Remote environment is used for monitoring and supervising appliances remotely. The system supports 3G and 4G internet, Wi-fi system and can be controlled by an android app. The user sends command from android app which gets transferred through internet. Security of the system is protected by firewall.

6.5 Configuration and Implementation of Smart Lighting System

There are two systems which have been proposed in the following manner. The title of the first system is, Self-Adjusting Lighting System and the other is Facial Recognition based Lighting Management System.

- **Self-Adjusting Lighting System**—This system is purposeful due to its design which helps save energy by analysing the current lighting intensity and accordingly controlling the artificial lighting. This brings automation to homes and saves time and energy which are both very important. Through the combination of smart programs and sensors, the user can control his/her home's lighting and even leave it up to the system to bring about energy saving without compromise on quality.
- **Facial Recognition Based Lighting Management System**—This system is designed to use Artificial Intelligence to bring about energy efficiency and make homes smarter. The system makes use of AI algorithms and hardware like cameras to create a lighting system which displays the power of computers and brings about energy saving. Through the combination of different sensors and devices, the user has complete control over his/her home's lighting.

Firstly, we will study the first system, i.e. Self-Adjusting Lighting System. We will understand the specifications of hardware, an architecture of this system, description on working of the system, workflow of the system.

A. Hardware Used

1. Arduino Uno
2. LM393 Optical Photosensitive LDR light sensor module
3. 5 V four channel relay expansion board
4. 9 W led bulbs
5. IR sensor

B. Connections for Lighting Control System

The male to male connector is connected from the 3.3v port of the Arduino to the positive railing of the breadboard which is then connected to one end of the LDR. A 10 K resistor connects the negative rail to the LDR. Another male to male connector connects the negative side of the LDR to the Analog I/O ports. The negative rail of the breadboard is connected to the ground port of the Arduino.

The 4-channel relay is connected to the arduino's Digital I/O ports using 4 male to male connectors which control the signals passed from Arduino to relay board.

For power, the relay board is connected to the breadboard's positive and negative railing using male to male connectors.

The relay board is connected to the 9 W Led bulbs through the wires from the wall or power source (Fig. 1).

C. **Working**

In the automatic mode, the arduino monitors the lighting intensity of the room using a LM393 Optical Photosensitive LDR light sensor module. It then passes the value to a function in the arduino which outputs a control signal to command the relay to switch on/off the bulbs to maintain a constant intensity.

It switches on/off individual bulbs depending on the lighting (natural + artificial) intensity of the room.

In the scenario where the combination of natural light and artificial light produces higher lighting intensity than required and without artificial light the lighting intensity becomes too less, the system will measure a difference between the desired intensity and the intensity produced by (artificial light + natural light) and natural light.

It will choose the option in which the difference between the desired intensity and the actual intensity is less.

The combination of the switched on bulbs creates enough intensity to maintain a predefined intensity set by the user during the installation of the system. To make the system easy to use, the user uses the GUI which lets the user try combinations of turned on bulbs and once he/she finds the comfortable lighting intensity, he/she can save that intensity in the app. The user can also control the lighting according to him/her from an android app or let the automatic system maintain the pre-set intensity (Fig. 2).

D. **General Information Graph**

The following section will explain in brief the various data the system will be dealing in with the use of graphs and tables. Two types of graphs are shown wherein the first graph shows the production of lumens by a Led bulb. The second and third graph show the difference in performance between the traditional and smart lighting system.

Lumens: Lumens is the amount of light emitted by a light source.
Lux: Lux is the amount of light falling on a specific area.

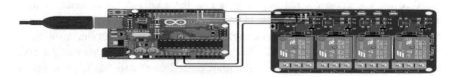

Fig. 1 Circuit diagram of self adjusting lighting system

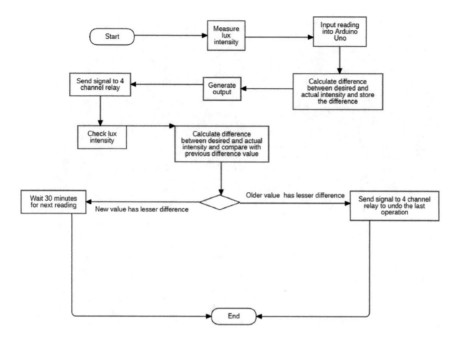

Fig. 2 Flowchart of self adjusting lighting system

The graph in Fig. 3 illustrates the amount of light produced in terms of lumens by a 9 W LED bulb. The X axis represents the number of 9 W LED bulbs and the Y axis represents the amount of lumens produced by the LED bulbs.

7 Traditional Lighting System

The graph in Fig. 4 explains how the lighting system works in the traditional (manual) lighting system. Here the user manually turns on and off the light bulbs. Sometimes the user may not turn off the light bulbs even when the natural light coming from outside is enough due to different reasons such as being busy or not being aware.

8 Smart Lighting System

The graph in Fig. 5 explains how the lighting is controlled by the smart lighting system. The system detects when the lighting intensity in the room is lower than the pre-set intensity selected by the user. It manages the number of bulbs turned on in order to produce enough lighting with respect to the natural light entering the room at the moment.

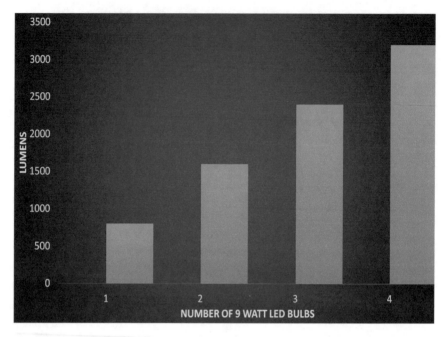

Fig. 3 Production of lumens by a Led bulb

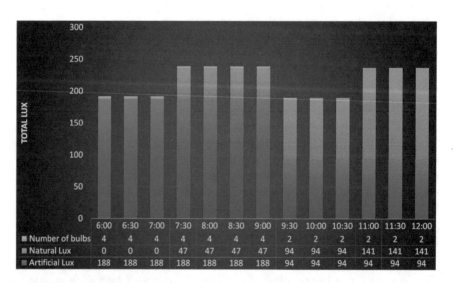

Fig. 4 Traditional lighting system

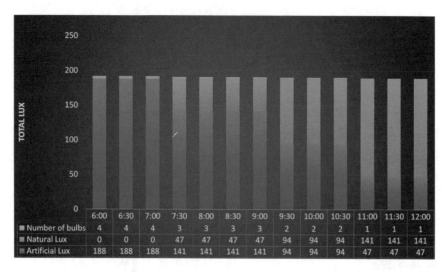

Fig. 5 Smart lighting system

9 Proposal of a Facial Recognition Based Lighting Management System

In this proposed system of facial recognition-based lighting management system, it has been assumed that it comprises of a camera and a data processing system. The camera runs facial recognition whenever a user enters the room. The user's face is mapped and stored in the database during the one-time setup and every user can set a predefined intensity as per his/her liking. Two IR sensors viz. sensor 1, sensor at a distance of 20 cm apart are installed in the doorway. When someone crosses the doorway, the sensors detect the movement and if sensor 1 detects movement before sensor 2, the system increments the count of the number of people in the room and the camera captures the person's face and checks whether that person is registered in the system. If the person is recognised, the system commands arduino to set the specific intensity. If the person is not recognised, the system sets the default intensity. If sensor 2 detects movement before sensor 1, it decrements the count of people in the room and if the count reaches 0, it will turn off the lights (Fig. 6).

The camera maps the room and looks for movement every 20 min. If no movement is found, it commands the system to dim the lights. The system keeps a timer of 20 min. The camera monitors the room for the slightest movement. If any movement is detected, it resets the timer, otherwise, when the timer finishes 20 min, the lights turn off.

The camera is connected to a computer where the facial recognition algorithm analyses the data and accordingly instructs the lighting system to set the intensity and control the lights.

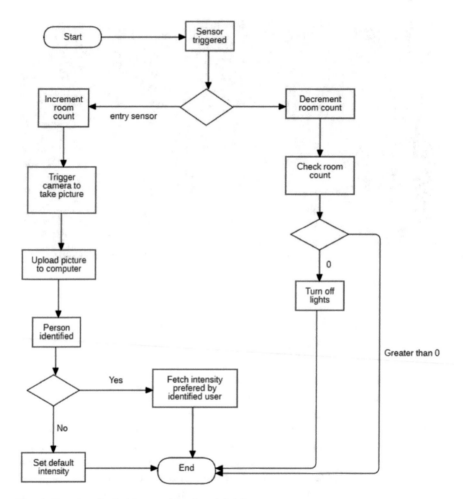

Fig. 6 Flowchart for facial recognition based lighting management system

10 Results for Self-adjusting Lighting System

The surface area mentioned for this experimental set up is 200 ft², is used as a reference in creating graph 4 and 5 depicted above.

Below is the calculation used in converting Lumens into lux where Lumens is the amount of light produced by a light source and lux is the amount of light falling on an area.

The illuminance Ev in lux (lx) is equal to 10.76391 times the luminous flux ΦV in lumens (lm) divided by the surface area A in square feet (ft²):

$$Ev(lx) = 10.76391 \times \Phi V \ (lm)/A\left(ft^2\right)$$

The above graphs 4 and 5 illustrate the amount of lux measured at different points of time and how many 9 W LED bulbs are needed to create the required lux.

The X axis shows the time from 6:00 am to 12:00 pm with a difference of 30 min.

The Y axis shows the lux measured by the LDR. The total lux is the sum of natural lighting lux and artificial lighting lux.

The blue section represents the lux intensity produced by the artificial light and the orange section represents the lux intensity produced by the natural light.

Electricity consumed by a 9 W LED bulb LED Bulbs power rating—9 W.

Quantity—1.

No. of hours a day—10 (Average usage).

Price of 1 kWh (1 unit) = Rs. 5

Hence, the total cost of running one 9 W Led bulb comes upto Rs. 0.045 per hour.

From the above data we can calculate the energy used by the LED bulbs in Graph 4 and Graph 5.

10.1 Calculations

Total cost = number of bulbs * number of hours * cost of electricity per bulb

Total bulbs: 4

Cost of running one 9 W Led bulb: Rs 0.0045/h.

Therefore, for graph 2 the calculation would be as below

Total cost = (4 * 4 * 0.045) + (3 * 2 * 0.045) + (2 * 0.5 * 0.045)

Total cost = Rs. 1.035

For graph 3, the calculation would be as below.

Total cost = (4 * 1.5 * 0.045) + (3 * 2 * 0.045) + (2 * 1.5 * 0.045) + (1 * 1.5 * 0.045)

Total cost = Rs. 0.7425

Hence with the smart system implementation, an average consumer saves Rs. 5734 per year in electricity bill (Table 2).

Table 2 Comparative approach of usage cost	Usage	Standard usage cost	Smart system usage cost
	Per hour	1.035	0.7425
	Per day	24.84	8.91
	Per month	745.2	267.3
	Per year	8942.4	3207.6

11 Conclusion

The future of expansion and development of smart lighting system is a multi-disciplinary research area. In this chapter, there is an exemplification of two techniques, one is Self-Adjusting Lighting System and the other is Facial Recognition based Lighting Management System. Both the approaches are the smart lighting systems. The main conclusions from the paper indicate that replacing CFL light sources with smart sensor-based systems improves energy saving.

However, it has been mentioned in the literature that multiple factors including the high start-up cost of these early smart lighting systems contributed to their low adoption rate. In fact, smart lighting systems was, and still many a times it is observed as a luxury purchase, with little to no benefit over traditional lighting technology.

But the authors of this chapter have proved that the smart lighting system as explained using both the approaches is beneficial and affordable too. With the smart lighting system, an average user can save more than 50% of the amount he/she pays for electricity. This not only saves money but also brings automation into homes which makes lives easier and simple and people may utilize the time for the other tasks to be carried out.

It is understood that the future of smart lighting is bright; however, its widespread adoption may get blocked as it does not justify the initial investment cost in many cases. This can be achieved using the development of novel control algorithms and thoughtful design of robust and feasible smart lighting frameworks.

References

1. Martirano, L.A.: Smart lighting control to save energy. In: Proceedings of the 6th IEEE International Conference on Intelligent Data Acquisition and Advanced Computing Systems, 15–17 Sept 2011, pp. 132–138 (2011)
2. Hsien-Tang, L.: Implementing smart homes with open-source solutions. Int. J. Smart Home 7 (4), 289–295 (2013)
3. Rath, D.K.: Arduino based: smart light control system. Int. J. Eng. Res. Gen. Sci. 4(2), 784–790 (2016)
4. Sripan, M., Lin, X., Petchlorlean, P., Ketcham, M.: Research and thinking of smart home technology. In: International Conference on Systems and Electronic Engineering (ICSEE'2012), December 18–19, 2012, Phuket, Thailand
5. Robles, R.J., Kim, T.: Applications, systems and methods in smart home technology: a review. Int. J. Adv. Sci. Technol. 15 (2010)
6. Redriksson, V.: What is a Smart Home or Building (2005, unpublish)
7. Sung, W.T., Lin, J.S.: Design and implementation of a smart LED lighting system using a self adaptive weighted data fusion algorithm. Sensors 13(12), 16915–16939 (2013)
8. Nurfarhana, N., Ilis, N.M., Juin, T.P.: Analysis of the light dependent resistor configuration for line tracking robot application. In: IEEE 7th International Colloquium on Signal Processing and its Applications (March, 2011)
9. Zoref, L., Bregman, D., Dori, D.: Networking mobile devices and computers in an intelligent home. Int. J. Smart Home 3(4), 15–22 (2009)

10. A. Standard Energy Efficient Design of New Buildings Except Low-Rise Residential Buildings. American Society of Heating, Refrigerating and Air-conditioning Engineers, Atlanta (1989)
11. Buchel, M.: Integration of technical systems in buildings using bus-technology—the new guideline 6015, VDI BERICHTE, 1639, pp. 9–14 (2002)
12. Von Neida, B., Manicria, D., Tweed, A.: An analysis of the energy and cost savings potential of occupancy sensors for commercial lighting systems. J. Illum. Eng. Soc. **30**(2), 111–125 (2001)
13. Guo, X., Tiller, D., Henze, G., Waters, C.: The performance of occupancy-based lighting control systems: a review Light. Res. Technol. **42**(4), 415–431 (2010)
14. Womack Electric Supply. http://www.womackelectric.com/wp-content/uploads/2011/05/P-S-Occupancy-and-Vacancy-Sensors-Catalog.pdf (2011)
15. Mowad, M.A.E.L., Fathy, A., Hafez, A.: Smart home automated control system using android application and microcontroller. Int. J. Sci. Eng. (2014)
16. Chew, I., Karunatilaka, D., Tan, C.P., Kalavally, V.: Smart lighting: the way (2017)
17. Tang, C.W., Huang, B.J., Ying, S.P.: Illumination and color control in red-green-blue light emitting diode. IEEE Trans. Power Electron. **29**(9), 4921–4937 (2014)
18. Higuera, J., Llenas, A., Carreras, J.: Trends in smart lighting for the Internet of Things. arXiv preprint arXiv: 1809.00986 (2018)
19. Tsai, C.C., et al.: Investigation of Ce:YAG doping effect on thermal aging for high-power phosphor-converted white-light-emitting diodes. IEEE Trans. Dev. Mat. Rel. **9**(3), 367–371 (2009)
20. Kim, Y., Charbiwala, Z., Singhania, A., Schmid, T., Srivastava, M.B.: SPORTLIGHT: Personal Natural Resource Consumption Profiler. HotEmNets 2008 (2008)
21. Waradkar, G., et al.: Automated room light controller with visitor counter. Imperial J. Interdisc. Res. **2**(4) (2016)
22. Jandial, A., Kumar, S., Butola, R., Pandey, M.K.: IR based home appliances control system. Int. J. Recent Innov. Trends Comput. Commun. **5**(5), 628–631 (2017)
23. Garg, V., Bansal, N.: Smart occupancy sensors to reduce energy consumption. Energy Build. **32**(1), 81–87 (2000)
24. Zhen, Z.-N., Jia, Q.-S., Song, C., Guan, X.: An indoor localization algorithm for lighting control using RFID. In: Energy 2030 Conference, 2008, ENERGY 2008, pp. 1–6. IEEE (2008)
25. Soheilian, M., Moadab, N.H., Fischl, G., Aries, M.B.: (2019, November)
26. Journal of Physics: Conference Series, vol. 1343, no. 1, p. 012155. IOP Publishing
27. Teymourzadeh, R., Ahmed, S.A., Chan, K.W., Hoong, M.V.: Smart GSM based home automation system. In: IEEE Conference on Systems, Process & Control (ICSPC), Kuala Lumpur, 2013, pp. 306–309 (2013)
28. Mahmud Rana, G.M.S.A., Mamun Khan, A., Hoque, M.N., Mitul, A. F.: Design and implementation of a GSM based remote home security and appliance control system. In: 2nd International Conference on Advances in Electrical Engineering (ICAEE), Dhaka, 2013, pp. 291–295 (2013)
29. Zhihua, S.: Design of smart home system based on ZigBee. In: International Conference on Robots & Intelligent System (ICRIS), Zhangjiajie, 2016, pp. 167–170 (2016)
30. Wenbo, Y., Quanyu, W., Zhenwei, G.: Smart home implementation based on Internet and WiFi technology. In: 34th Chinese Control Conference (CCC), Hangzhou, 2015, pp. 9072–9077 (2015)
31. Malche, T., Maheshwary, P.: Internet of things (IoT) for building smart home system. In: International Conference on I-SMAC (IoT in Social, Mobile, Analytics and Cloud) (I-SMAC), Palladam, 2017, pp. 65–70 (2017)
32. Piyare, R.: Internet of things: ubiquitous home control and monitoring system using android based smart phone. Int. J. Internet Things **2**(1), 5–11 (2013)
33. Batool, A., Rauf, S., Zia, T., Siddiqui, T., Shamsi, J.A., Syed, T.Q., Khan, A.U.: Facilitating gesture-based actions for a Smart Home concept. In: International Conference on Open Source Systems & Technologies, pp. 6–12. IEEE (2014, December)

QUIC Protocol Based Monitoring Probes for Network Devices Monitor and Alerts

Anurag Sharma and Deepali Kamthania

Abstract In the coming days, 5G networking needs faster software automation in the existing environment. With this view in this paper, an attempt has been made to formulate an approach for enhancing the HTTP based monitoring without affecting the current services. The automation can be performed using modern web solutions. A node is considered as the smallest host inside an intranet and internet network based on the physical or logical grouping of multiple networks. It acts as host computers when identified by an IP address (ipv4 or ipv6) and when connected with the host to many clients it is identified through its network subnets. The communication among these nodes can be improved by using UDP based HTTP3.0 along with TLS 1.3 for GET or POST requests through UDP streams. The traditional HTTP stack ordinary monitoring can be upgraded through QUIC Protocol for a faster and more efficient approach in future networks and real-time monitoring. In the QUIC based HTTPS scenario, it has been observed that the load time takes less than 200 ms in network latency, which results in a faster approach as compared to handshake between host to client and vice versa during the previous HTTP introduced approaches. The slow response time results in wait, which causes a penalty. The suggested approach can be beneficial at a network node for monitoring the connected nodes in the network by sending QUIC protocol-based transport layer beacons in certain time-lapse, resulting better and faster alerting in information technology infrastructures.

Keywords HTTP (hyper text transfer protocol) · Domain name system (DNS) · Quick UDP internet connections (QUIC) · Universal stream transport protocol · Multi-stream protocol · Latency reduction · Multi-layer transport protocol

A. Sharma · D. Kamthania (✉)
School of Information Technology, Vivekananda Institute of Professional Studies,
Delhi, India

© The Author(s), under exclusive license to Springer Nature Switzerland AG 2022
U. Singh et al. (eds.), *Smart Sensor Networks*, Studies in Big Data 92,
https://doi.org/10.1007/978-3-030-77214-7_6

1 Introduction

In, today's scenario internet has basic necessity and the amount of data is growing consistently with each millisecond and need to be transferred in a fast, reliable and secure manner using present HTTPs stack infrastructure. The HTTP is transported over TCP and is secured through TLS but there is scope for improvement in the connection and data transfer between client and server using Quick UDP Internet Connections (QUIC) developed by Google instead of the traditional Internet stack at both ends. Quic combines transport and security layer into one violating OSI model and has improved features over TCP/TLS. To ensure reliable and secure communication between two ends it is essential to have transport and security protocols TCP/TLS and UDP/DTLS but there are connection overhead, latency, and connection migration issues when used in some applications [1]. In order to communicate using QUIC implementation need to be done at both client and server end. Therefore, QUIC client support is required at application level, like in the browser [2]. In this paper we propose an approach using HTTP and QUIC protocol for improving the response time for modern network interconnectivity. The proposed method provides response in intranet with the latest software faster real time available open sources under one second.

2 Background

2.1 TCP and UDP

The transport layer supports segment exchange in end to end interacting applications, generally TCP (Transport Layer Protocol) and UDP (User Datagram Protocol) are used. To avoid buffer overflow at the receiver, a reliable end-to-end connection is provided by TCP through congestion control mechanism. Many TCP version have been proposed for addressing the demand for throughput increase resulting in connection establishment and resource (i.e., processor, memory, energy) utilization overheads [3, 4]. UDP have lower overheads as compared to TCP as it does not provide the said features. The TCP handshake highly affect the connection start-up latency, 1 Round-Trip Time (RTT) is required for TCP and 2 or 3 RTTs are required when TLS (Transport Layer Security) is added to protocol [5] which further increases in case of unreliable wireless link and cause frequent connection drops [6]. There is wastage of resources in case high connection is established in such case for transferring small amount of data. This issue is resolved by TCP Fast Open piggybacking data in SYN segments [7] but it is not scalable [8]. Any change in network parameters (such as IP address or port) breaks the connection so connection need to be re-established, or data flow is rerouted using a gateway [9]. In case of reboot or device crash half open connection [10] can result in SYN flooding [11, 12] and for head-of-line blocking the receiver has to wait for drop packet retransmission [13–15]. The limitations of UDP and TCP, shows that there is need for the enhancement of transport layer protocols for performance improvement. QUIC is a protocol for user space, UDP-based, stream-based and multiplexed transport generated by Google. According to [8], around 7% of the world-wide Internet traffic employs QUIC. This protocol offers all the features needed to be regarded as a protocol for connection-oriented transport. In addition, QUIC solves the numerous problems faced by other connection-oriented protocols such as TCP and SCTP [16].

2.2 TLS and DTLS

Along with reliability secure data transmission is also required over transport layer. Most common connection-oriented and stateful client-server cryptographic protocol is TLS [17]. TLS handshake is unique per connection and have symmetric encryption, between client and server two round trips are required. The overheads are imposed in case of there is connection drop due to sleep phases, connection migration, and packet loss which can be resolved using a lighter version of TLS for datagrams, named DTLS (Datagram Transport Layer Security) [18] which does not require a reliable transport protocol as it can encrypt or decrypt out-of-order packets so it can be used with UDP. DTLS is more suitable for resource-constrained devices communicating through an unreliable channel. TLS and DTLS has been designed for point-to-point communication and face challenge to secure one-to-many connections such as broadcasting and multicasting. DTLS does not support connection migration it identifies connections based on source IP and port number [19] and DTLS handshake packets are large and may fragment each datagram into several DTLS records where each record is fit into an IP datagram causing overheads [20].

2.3 QUIC

QUIC employs some of the basic mechanisms of TCP and TLS, while keeping UDP as its underlying transport layer protocol [1]. QUIC is the emerging transport layer protocol, providing encrypted, stream-multiplexed, low-latency data transfer to improve web performance for HTTPS [8]. With the recently initiated IETF standardization [21] QUIC, which started as a TCP replacement to transport HTTP/2, is becoming a universal transport protocol [22]. QUIC is a combination of transport and security protocols performing encryption, packet re-ordering, and retransmission [1]. Viernickel et al. [23] proposed a multipath-enabled QUIC (MPQUIC) to leverage multiple network interfaces on mobile devices which increases throughput as compared to traditional QUIC, TCP and multipath transport protocol MPTCP.

In today's scenario QUIC with multipath support feature is becoming a universal stream transport protocol for internet. The deployabilty issues are resolved as QUIC works on top of UDP which enables its implementation without modifying operating systems resulting in easy upgrade and shipment with applications. encryption and authentication are always provided except for handshake and reset packets. In QUIC in a single connection several requests are handled simultaneously using multiple streams having StreamID's avoiding blockage of entire connection [8, 24]. In contrast to TCP, the QUIC connection is identified by the ConnectionID to enable connection migrations between IP-address/port tuples rather than IPaddress/ port tuple [23].

Figure 1c shows that an authenticated header and encrypted payload frames of multiple streams are available in QUIC packet. QUIC implementation is easily upgradable and can be easily shipped along with applications as deployabilty issues are resolved because QUIC work on top of UDP. The middlebox manipulation are prevented in QUIC operations and limits protocol conformity, except for handshake and reset packets encryption and authentication are always provided [8, 24]. Figure 1c shows that QUIC payload is always encrypted. In one handshake cryptographic information are shared with transport information to have low latency. Once the client has knowledge of server initial key, the new connection can be established with zero round trip time (0-RTT) data delay.QUIC offers stream-multiplexing, which is the ability to use multiple streams to manage several request/response pairs simultaneously on a single link using frames internally. The multiple frames of different streams and additional control frames is included in a QUIC packetEach stream is recognized by a StreamID and has its own Offset-header sequence number space to minimize inter-stream dependencies (Fig. 1c). Packet loss therefore does not block the entire link, but rather the streams that are transmitted in the missing packet that are affected preventing head-of-line blocking between streams. A client or a server sends data, marked with an unused StreamID, to open a stream, indirectly causes a new stream on the receiving side to be formed resulting in a 0-RTT stream establishment.

Figure 2 shows the QUIC connection establishment between client and server. An inchoate hello (CHLO) message is sent by client to the server and the server answers with a reject (REJ) containing a signed server configuration, a certificate chain and a source address token [8] on the basis of which the client sends a complete CHLO message, containing its ephemeral cryptographic information which is used for fast connection establishment with a 0-RTT latency. The server answers with a server hello (SHLO) message containing own ephemeral cryptographic information which completes the handshake based on which both sides calculate keys for encrypting subsequent data [8]. Unlike TCP, the IPaddress/port tuple does not define the QUIC connection, but uses the ConnectionID to allow connections between IP address/port tuples [23].

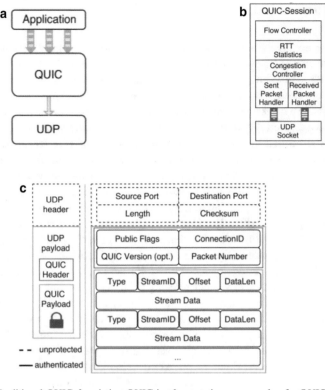

Fig. 1 a Traditional QUIC, b existing QUIC implementation, c example of a QUIC packet [21]

Fig. 2 Initial QUIC
connection establishment [23]

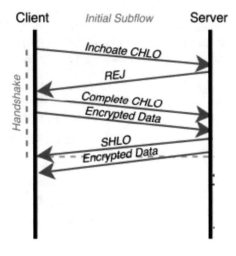

3   Traditional Network and QUIC

Most of the applications are based on the Transmission Control Protocol (TCP). In TCP the packets are organized and data transport is reliable the lost data is identified and delivered. The congestion and flow control avoid network overload. The transport layer features take care data availability making application level protocol (HTTP) hassle free. In the traditional web stack, to transport HTTP data TCP protocol is used. To refer to the OSI model, transferring HTTP over TCP is additionally protected by TLS, a security setup on top of TCP (compare Fig. 3).

HTTP/1.1 opens several TCP connections to fetch data quickly and in parallel (as shown in Fig. 4a). However, this strategy can be inefficient and may cost high processor rates on limited devices because each single link has to be treated. HTTP/2.0 suggested that a single, but multiple streams using TCP link. Data can be supplied on any stream (Fig. 4b). Due to TCP's in order delivery, the issue of this method is the head of-line-blocking delay. If one stream's packet is lost, then all other streams are blocked. The head of the line is blocking the connection as a whole. QUIC makes some streams, such as HTTP/2.0, but because of a blocking stream, it does not block any other streams (Fig. 4c). Data transmission on other streams is not delayed because the in-order delivery is not bound by UDP. Streams are defined by a specific stream ID according to the latest core protocol draft, and data distribution ordering is done with stream offsets within a stream [2].

Although TCP is one of the widely used transport protocol transmits data but it has several disadvantages. Having a quicker development cycle and publishing new versions of kernel-based implementations such as TCP is more difficult. TCP-based network connections run in kernel mode. In both the client and the server side, computers also have to be upgraded. In turn, introducing changes to TCP and thus to the kernel will cause operating system changes. In addition, TCP has such inconveniences, such as the latency of head-of-line blocking and handshake delay. At least one round trip is required for the transfer of data to set up the TCP link. In addition, the TLS security layer adds two additional round trip delays to top TCP

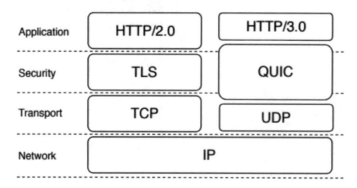

Fig. 3   Traditional network stack and QUIC in comparison, adapted from [8]

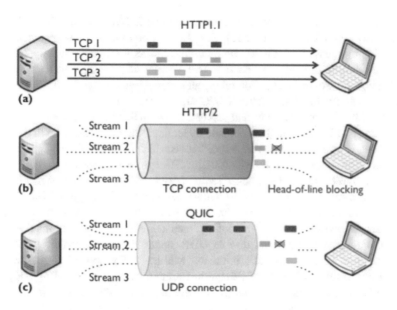

Fig. 4 The different HTTP versions and the upcoming HTTP/3.0 in comparison [25]

connections (compare Fig. 1). Even though with TLS 1.3 and TCP Quick Accessible, the handshake delay seems to be solved, the data transfer can still be optimized. A new strategy exists to decrease handshake latency. In comparison to TCP, QUIC operates in the user space and uses the User Datagram Protocol (UDP) as the underlying transport protocol. UDP is a transportation protocol commonly used and lightweight. It is also ideal to use data transfer from host to host as a transport layer protocol. The benefit of UDP is that like firewalls, it can traverse middleboxes. Many firewalls block unknown protocols, especially in large firms. Thus, the middle boxes could drop unfamiliar packets. QUIC encrypts its packets and authenticates them and makes it possible to transport them via UDP. There is a greater chance of using the Internet to get UDP packets. However, QUIC has to implement everything that makes the protocol safe, stable, fast and reliable on its own because of the necessary but missing TCP features in UDP: QUIC has to implement congestion control and flow control. The characteristics like handshake, algorithm for encryption and authentication are implemented in the application layer by QUIC. TCP is not used because of its slow development time. This makes it simpler to settle on the UDP protocol for data transport. It is also more convenient to add new releases and adapt QUIC features without disrupting the long-term development cycle of the TCP based kernel. TCP's involves three-way handshake, on the top of TCP, TLS 1.2 is used for encryption and authentication requiring two trips along with one trip for data. For QUIC, TLS 1.3 is used as encryption and transport are in same layer in user space [2]. TLS 1.3 is already an IETF standard described in RFC 8446 but hardly implemented but have high potential to become standard security protocol in future. OpenSSL package in Ubuntu was equipped with TLS 1.3. TLS 1.3 will be used by more applications in next Ubuntu release [26]. QUIC can be used by users through Google applications like the Chrome browser, the open source version Chromium, and the YouTube application on Android [1]. For client and server interaction QUIC need to be implemented on the both. After QUIC standardization by IETF QUIC can be supported by different browser and servers. If the QUIC flag is allowed in Chrome and a request via TCP and TLS is made by a client, the QUIC server will display a QUIC flag in its HTTP response. A race between TCP/TLS and QUIC [1] would be the next client-side order, if the client wishes. The protocol stack which will be used for that request will be the quicker reply. QUIC will only be chosen if QUIC is allowed and supported in the entire client-to-server path. To trigger QUIC, Chrome and Chromium users can currently set the corresponding QUIC flag in their browsers. Because of UDP spoofing attacks, servers and companies with production-based services can disable UDP inputs via their firewalls. QUIC will currently not be used on such intranets. Multipath Transport Control Protocol (MPTCP), Multipath Fast User Datagram Protocol Internet Link (MP-QUIC) and Stream Control Transport Protocol (SCTP) are used for multi-connectivity transport layer solutions but have low middlebox compatibility [27]. MP-TCP [28] is a multipath protocol that uses several TCP connections that according to various schemes, carry traffic but lack prioritization which limits the scope [29]. The Redundant Scheduler can be used to send duplicated packets to all available TCP connections for improved reliability [30]. The multiple QUIC connections [31] are used by MPQUIC [32]. Multi-stream transport protocols such as SCTP and QUIC resolve the problems faced by TCP, such as Head of Line (HoL). For multi-stream transport protocols, an SDN-based architecture Networks of multipaths have been proposed. The proposed framework provides API framework for multi-stream rules to define applications for SDN services [33]. QUIC is a reliable UDP-based transport protocol which provides TCP with the extension of data multiplexing and prioritization with similar functionality. Although still in the early stages of development, MP-QUIC offers a structure inherited from QUIC that allows multiple application data sources and prioritization of these sources, but does not support multi-path redundancy [32, 34]. A new selective redundant MP-QUIC concept was proposed to provide low latency and full redundancy for prioritization of application streams and high aggregated bandwidth for background application streams in the same connection [35].

4   HTTP Based Monitoring Systems

HTTP based monitoring systems considers various intermediate services in client server architecture to deliver web pages through various web servers [36] using GET/Post method with request status code like 2xx/4xx/5xx and handle various redirects [37]. On the basis of physical or logical grouping of multiple networks the node is considered as smallest host to internet or intranet, it acts as host identified by an IP address (ipv4 or ipv6) and can be connected to many clients through network subnet. It acts between client and server to deliver webpages. A web server acts as a handler and serve client request to deliver web pages over IT infrastructure (Public/Private) network [38]. Therefore, when a client end web browser requests a web page, it is delivered through OSI/TCP IP model-based layer 7 with the socket port 80 HTTP and 443 for HTTPS. During the communications the specific protocols plays their roles to deliver optimum web browser-based content from server to client network using DNS (Domain Name System) requests for translating domain name to I.P. address and vice versa with reverse name lookup which is performed internally by the client browser [39]. Most of the time in web related work HTTP is used as a service to perform various to perform data sharing known as MIME [40]. HTTP is a simple one-way monitoring for website/intranet IP address as a service host monitor for checking whether the destination TCP ports i.e. 80 and 443 HTTPS service is active or not. This can also be implemented using HTTP REST API or Gateway solutions. HTTP can also be used as services in cloud computing and load balancing for content delivering networks (CDN) like CloudFlare, Aws cloudfront. It can also be used for authentication [41]. HTTP can also implement as in our devices for getting the live Realtime updates in in case manufacturer have provided service as HTTP Server for the devices real-time update. HTTP Service is quite slow for ordinary versions like HTTP 1.0/HTTP 1.1 or HTTP 2.0 so the handshake process in TCP model takes time [42]. The proper time data delivering and smart sensors devices data streams can be performed with using QUIC protocol once the protocol is effective in real world implementation through wireless propagation over 5G as well intranet-based networks [43].

4.1 Host Monitoring Services

In this section the basic HTTP mechanism is shown by using WhatsUP GOLD version 8.0 from IPSWITCH Tool is used for the simulation [44]. A simple host monitor scan is performed to check up and down services for port TCP 80.

Figure 5 shows HTTP monitoring as in HOST performed on local host setup with IP Address 127.0.0.1 and the green background shows the active state. Suppose the HTTP Service is down due to some reasons the webserver is not able to act to request from the source then the host monitor alert the poll request and change the general host monitors to an alert message i.e. red in colour as shown in Fig. 6. The free available Wireshark tool for packet capture in details over local loopback localhost tells how this host monitor concept work in terms of the sending request and responding over it as shown in Fig. 7.

Figure 8 shows the live packets captured in Wireshark when the packet is not able to respond to the server then the re-acknowledgement send repeatedly.

In case the packet from source to destination is favourable then the host monitor is showing active response for HTTP based alert as shown in Fig. 9 inside wire

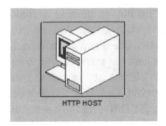

Fig. 5 Active HOST

Fig. 6 Down HOST

```
◢ Wireshark · Packet 1 · Adapter for loopback traffic capture
>  Null/Loopback
∨  Internet Protocol Version 4, Src: 127.0.0.1, Dst: 127.0.0.1
       0100 .... = Version: 4
       .... 0101 = Header Length: 20 bytes (5)
     > Differentiated Services Field: 0x00 (DSCP: CS0, ECN: Not-ECT)
       Total Length: 52
       Identification: 0x7d7b (32123)
     > Flags: 0x4000, Don't fragment
       Fragment offset: 0
       Time to live: 128
       Protocol: TCP (6)
       Header checksum: 0x0000 [validation disabled]
       [Header checksum status: Unverified]
       Source: 127.0.0.1
       Destination: 127.0.0.1
>  Transmission Control Protocol, Src Port: 49709, Dst Port: 80, Seq: 0, Len: 0
```

Fig. 7 Wireshark packet from source to destination consist of the packet data

2 0.000019	127.0.0.1	127.0.0.1	TCP	44 80 → 49709 [RST, ACK] Seq=1 Ack=1 Win=0 Len=0
3 0.500510	127.0.0.1	127.0.0.1	TCP	56 [TCP Retransmission] 49709 → 80 [SYN] Seq=0 Win=65535 Len=0 MSS=65495 WS=256 SACK_PERM=1
4 0.500541	127.0.0.1	127.0.0.1	TCP	44 80 → 49709 [RST, ACK] Seq=1 Ack=1 Win=0 Len=0
5 1.001463	127.0.0.1	127.0.0.1	TCP	56 [TCP Retransmission] 49709 → 80 [SYN] Seq=0 Win=65535 Len=0 MSS=65495 WS=256 SACK_PERM=1
6 1.001478	127.0.0.1	127.0.0.1	TCP	44 80 → 49709 [RST, ACK] Seq=1 Ack=1 Win=0 Len=0
7 1.502116	127.0.0.1	127.0.0.1	TCP	56 [TCP Retransmission] 49709 → 80 [SYN] Seq=0 Win=65535 Len=0 MSS=65495 WS=256 SACK_PERM=1
8 1.502199	127.0.0.1	127.0.0.1	TCP	44 80 → 49709 [RST, ACK] Seq=1 Ack=1 Win=0 Len=0
9 2.016351	127.0.0.1	127.0.0.1	TCP	56 [TCP Retransmission] 49709 → 80 [SYN] Seq=0 Win=65535 Len=0 MSS=65495 WS=256 SACK_PERM=1
10 2.016367	127.0.0.1	127.0.0.1	TCP	44 80 → 49709 [RST, ACK] Seq=1 Ack=1 Win=0 Len=0

Fig. 8 Wireshark packet response when HTTP destination service is not working

```
>  Transmission Control Protocol, Src Port: 49710, Dst Port: 80, Seq: 1, Ack: 1, Len: 62
∨  Hypertext Transfer Protocol
   ∨  HEAD / HTTP/1.0\r\n
      >  [Expert Info (Chat/Sequence): HEAD / HTTP/1.0\r\n]
         Request Method: HEAD
         Request URI: /
         Request Version: HTTP/1.0
      Accept: */*\r\n
      User-Agent: WhatsUp_Gold/7.0\r\n
      \r\n
      [HTTP request 1/1]
      [Response in frame: 6]
```

Fig. 9 Wireshark freeware tool reference a simple view of HTTP packet capture from sender side on loopback interface

shark packet tracing method for loopback adaptor for general host monitoring (Fig. 10).

Once the packet is able to communicate when server is answering the request it can re-acknowledge the same and will be shown like the below live packets captured in Wireshark Fig. 11.

```
>  Transmission Control Protocol, Src Port: 80, Dst Port: 49710, Seq: 1, Ack: 63, Len: 244
∨  Hypertext Transfer Protocol
   ∨  HTTP/1.1 200 OK\r\n
      >  [Expert Info (Chat/Sequence): HTTP/1.1 200 OK\r\n]
         Response Version: HTTP/1.1
         Status Code: 200
         [Status Code Description: OK]
         Response Phrase: OK
   >  Content-Length: 696\r\n
      Content-Type: text/html\r\n
      Last-Modified: Sat, 26 Sep 2020 17:27:14 GMT\r\n
      Accept-Ranges: bytes\r\n
```

Fig. 10 Another Wireshark freeware tool reference a simple view of HTTP packet capture at receiver side loopback interface

1 0.000000	127.0.0.1	127.0.0.1	TCP	56 49710 → 80 [SYN] Seq=0 Win=65535 Len=0 MSS=65495 WS=256 SACK_PERM=1
2 0.000047	127.0.0.1	127.0.0.1	TCP	56 80 → 49710 [SYN, ACK] Seq=0 Ack=1 Win=65535 Len=0 MSS=65495 WS=256 SACK_PERM=1
3 0.000080	127.0.0.1	127.0.0.1	TCP	44 49710 → 80 [ACK] Seq=1 Ack=1 Win=2619648 Len=0
4 0.003732	127.0.0.1	127.0.0.1	HTTP	106 HEAD / HTTP/1.0
5 0.003749	127.0.0.1	127.0.0.1	TCP	44 80 → 49710 [ACK] Seq=1 Ack=63 Win=2619648 Len=0
6 0.445719	127.0.0.1	127.0.0.1	HTTP	288 HTTP/1.1 200 OK
7 0.445738	127.0.0.1	127.0.0.1	TCP	44 49710 → 80 [ACK] Seq=63 Ack=246 Win=2619392 Len=0
8 0.561131	127.0.0.1	127.0.0.1	TCP	44 49710 → 80 [FIN, ACK] Seq=63 Ack=246 Win=2619392 Len=0
9 0.561196	127.0.0.1	127.0.0.1	TCP	44 80 → 49710 [ACK] Seq=246 Ack=64 Win=2619648 Len=0

Fig. 11 Wireshark packet response when HTTP destination service is working and serving status response code 200

Figure 12 shows earlier versions of HTTP/HTTPS based scenario and working. The communication between client and server takes place with requests through handshaking method. This is a time-consuming process so ICMP better option as compare to HTTP based service.

In the current scenario on live real-time network reachability generally ICMP based monitoring is used for gopher protocol. It is different from transport protocols such as TCP and UDP. It acts as a supporting protocol in IP Suite used in network devices. IP suite use version 4 and 6 uses ICMP which involves transmission through datagram having an IP header encapsulating the ICMP data [45]. It consists of codes values inside IP address packet codes (Type 0 with code 0 for Echo reply) usually called PING and (Type 3 with code 0 named as destination network unreachable, or code 1 for destination host unreachable). Similarly type 0 to 255 act as control message for monitoring using PING. In ICMP redirects demands occurs by a message passing information by a client, forward to switches/routers and pass to directing data towards defined gateway [46]. In case the time period for transmission exceed the time limit in datagram is neglected. The time surpassed messages can be tracked by the trace route utility to recognize entry ways or the ways between two hosts.

CURL3 QUIC based can be used in place of ICMP/Ping in these proposed probes for monitoring and software's can be developed on the proposed said architecture using MIME "X-Content-Type-Options" using custom defined hard-coded software's implementation. This can be implemented by adopting hard core socket connection injecting and so monitoring can be performed.

QUIC protocol performs the HTTP delivery through a bunch of multiplexed streamed UDP based data in forms of streams, so this approach is beneficial in terms of time taken, delay elimination and reduces the drawbacks of HTTP and HTTPS based versions like 1.0/1.1 and 2.0. Figure 8 shown simple TCP HTTP response when using Transport Later Security (TLS) in terms of old versions consuming more time and results in slow response for the time to await response by using QUIC. Figure 13 shows that this can be improved not only with smart sensor devices but with faster automations.

Fig. 12 Image source Cloudflare simple HTTP response source Cloudflare blog

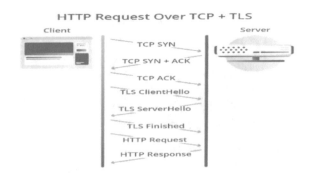

Fig. 13 QUIC protocol based HTTP response source Cloudflare blog

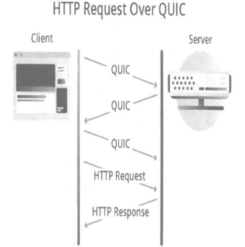

5 Need for QUIC in Existing Infra Structure

Figure 14 show the architecture for proposed QUIC based monitoring which involves QUIC—HTTP 3.0 probe as virtual engine services. The probe is an identification name of a device that act as a host to deliver monitoring services that proxies under network port internally/locally or can be externally through co-locations points encrypted with TLS (Transport Layer Security) in public cloud/ intranet or with inside based droplets over cloud or locally with physically infrastructure devices that proposed to adopt without affecting present architecture. HTTP acts as a GUI Tool implementation in virtualized environment for HTTP3 servers and inter message passing from host to client and host to host with faster message passing vice versa internally and updating through lightning fast speed with real time response as per diagram and bi-directional communicate with Upstream. Slave Probes works actively inside the monitoring cluster. Failover through upstream are backend behind NGINX based web server required for containerized or virtualized through Docker Images/minikubes for virtually creating the monitor probe communicating with each. In contrast, in case of a host down this replicate the upstream of the load balance for the monitor probes in disturbed infrastructure/cloud. Alternatively, clients can also install HTTPS based QUIC protocol enabled service for an identical form that it can differentiate between host/ clients. The proxy with current requests is understandable by the current infrastructure through proper response and learn to deal with the QUIC based requests and actively respond to them.

Prototype comprise minimum one device and maximum two devices that can use as a host and alternatively backup load balanced slave host for sending probes that actively responding through the inter connectivity basis. Some advance technical understandings are highly required while building this proposed architecture, and

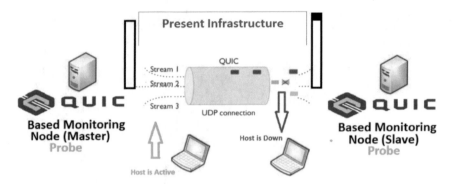

Fig. 14 Proposed architecture of QUIC based monitoring with host probe and slave node as backup for disaster recovery system

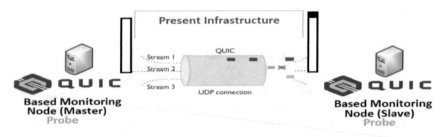

Fig. 15 QUIC Protocol based probe demonstration as a host node with slave node actively working

most of the time it can be shown unsuccessful results because of resources not available while using this new challenging statements. Some prototype conflicts on results shown, but enhancing the new adoption of resources could be better for the architecture. The approach has been proposed for monitoring and altering because of the following reasons:

i. **Hardware**

The proposed approach can be implemented on a dedicated device preferably Linux. The hardware requires services installation with GPL licences such like GitHub, etc. The minimum hardware requirement only for the simple CPU/Memory and active network enable device, Ordinary (LAN) or wireless service both dual band frequencies 2.4 and 5.0 GHz can be used. So, hardware plays a big role in terms of devices intercommunication and most important an active robust power supply over the device is used in host devices and timely maintenance is required for the hardware to perform tasks. A dedicated hardware virtual cloud machine can be used.

ii. **Probe**

Monitoring host node is called as "probe" for sending QUIC protocol-based data streams through its learning defined for client nodes, it actively sends the streams with no delay and constantly update in real time without affecting the other working operational architecture in devices while working simultaneously. Probe is a beacon or can say a stream that is going to send some data streams hard-coded by cron jobs (in Linux) or any methods sent within time intervals. The probe usually sends actively streams of data through network without the collision because of QUIC Protocol properties. Some multiple protocol over QUIC are still in the development phase and are not effective in the foresaid approach in probes. Figure 12 showing the Multiple Streams and actively UDP based QUIC protocol-based probe actively sending.

iii. **Network Architecture**

It requires the networking architecture as per proposed architecture shown in Fig. 14. If the subnet is connected in class C with cisco inter domain routing notation CIRD (/24 i.e. 255.255.255.0). The subnet can only perform monitoring up to 253 nodes due to restriction of /24 CIDR notation-based subnets, so a highly super subnet class IP address schema is required for the proposed node to perform big data network handling. This probe installation is mandatory over the core end only inside the network infrastructure where all the devices simultaneously contact from core end and the responses that can easily be accessed by the probes with no challenges for delivering the data set. It can perform the task with no issue on it if the IPV6 is also a good option in terms of "IPV4 address was in limited numbers stage" in public static IP address as per IANA Network architecture easily establish by highlighting the points over network core layer to distribute and further moving to access layer in IT infrastructure as well not needed in case of small network intranet premises.

iv. **Software**

Software is the basic backbone of the proposed approach. Software plays a major role like some real time updates can only be implemented by software and also by the latest languages for software beneficial and for faster update on the upcoming days like Realtime java scripts (Angular JS/NodeJS) could be plays crucial role for getting response alerts on the defined machines monitoring and alerting in the GUI environment. Recently, QUICK.CLOUD is beta available for using WordPress based websites when using Lightspeed cache plugin and delivered the websites QUIC based speed improvements for modern web. Further, if such a start is available in these days, so we can further move towards the development in this protocol using for software related approaches.

v. **Virtualization**

Virtualization here also can play a role in this proposed solution. By using virtual machine, we can implement it through free tools by (Oracle Corporation)

VirtualBox and VMware (Licensed). This can be implanted by adopting the defined software packages for alerting and monitoring prospects. The virtual machine also acts in the host for sending the probes from bi-directional communication and further can be extended for updating without affecting the current system. In terms of smart devices, it can only be possible when device having its operating system supports virtualization (only supporting the software to accept QUIC protocol probes responding). Virtualization created a big picture here when delivering the extensions for further updates over it. Some services like microKubes can be beneficial when using a guest operating system and install all these to perform.

vi. **Extensibility (API)**

API once the proposed architecture will build then moving further it can also be implemented on extension by various SAAS (Software as a Service) Cloud Computing model, this resulting a slight down a step on cost reduction in building multiple isolated environments and increasing the distributed approaches throughout which is a good option. Further implementing API such as REST (using JSON based) will be beneficial if proper development might be on track by software building procedures. Some more extensibility for upcoming days illustrated the more advance adoption in this proposed architecture and using by QUIC protocol is currently in draft by IETF/IEEE. Smart Sensors devices will be beneficial for the same and once this implemented then faster extensibility will be applied according to the technology. Moving further the CURL3.0 sending the same and using automations like Machine Learning approaches for the collected data can be easily possible for visualizing the active uptime reports, internal response time data for collecting results in big data terms and will be share for further performance improvements. Smart devices are costly with maintaining the continuity, and by upcoming years this extensibility further enhanced throughout the scope enhancements.

5.1   Handshake Challenges for Inside Conflict Environment

There are security threats in the network. Flood packets through streams can stop services of the host machine. DDos Attacks, can harm the system or will create higher computation power/service unavailable in the real world. These attacks can be reduced by using secured customized socket number. QUIC distinguishes between the 1-RTT and the 0-RTT handshake [47]. The cryptographic handshake minimizes the handshake latency by using known server credentials on repeat connections. A client can store information about the server and can use the 0-RTT handshake on subsequent connections to the same origin. The 1-RTT handshake is possible because the transport and cryptographic keys are overlapped into the same layer. The 0-RTT handshake is possible, since TLS 1.3 provides a 0-RTT and TLS 1.3 will be used as a security layer in QUIC. As many network connections are to the same servers which were contacted before, the 0-RTT makes it possible for a

Fig. 16 Cloudflare reference for HTTPS based timeline graph for QUIC

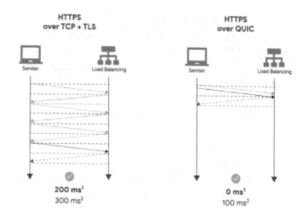

client to send payload without repeating cryptographic or transport key exchange. This resulted in control over repeated processes [48]. NAT (Network Address Translation) is proposed as breath taker for QUIC [49].

Figure 16 shows HTTPS based QUIC to QUIC communications over custom admin defined ports at application layer. The sockets in UDP streams are defined by the administrator while installing monitoring probes this allows POST HTTP/HTTPS query and listens only to the other side QUIC based devices. This is best practice for a small IT company using nodes and one or two servers to handle requests from many clients with their different IP addresses. In case of intranet for an active enterprises end solution it will be directly under the company network, application layer uses client spouse using NAT/PAT (Port Address Translation) inside their public IP through port forwarding application layer server end. According to Fig. 15 server block of HTTPS inside the HTTP web application and use multiple URL responses and handling. This results in faster sharing responses time as compared to ordinary HTTP/ICMP based. The QUIC based UDP streams are not allowed unless a proper kernel is built. The proposed system works fine for small organization but for large organization the implementation cost increases as smart sensors need to be injected in the software over kernels inside devices. As we are using network in terms of intranet it is okay but when it goes beyond the intranet, the global internet requires the local government approvals from Computer Emergency Response Teams (CERT's) teams to allow the traffic to be encrypted with VPN (self-build).

6   Experimental Setup and Evaluation

A. Server Side Setup: Ubuntu 20.04.1 (Linux 5.4 kernel and Mesa 20.0.8l) Virtual Machine (VM) in VirtualBox hosted the web server. It was allocated 4 processors and 8 GB of memory. Cloudflare's QUIC, public repository and available QUICHE

Library that support HTTP/3 and TLSv1.3 for supporting NGINX v1.18 web server. H3 support for HTTP connections to the server to clients was in the alt-svc header. TLSv1.3 handshaking was used on both H3 and H2 over TCP + TLS connections and CUBIC congestion management was used. TCP tuning and parameterization out-of-the-box were used. Cloudflare states that NGINX does not officially support their H3 fix. Open LiteSpeed based was used as a web server.

Client-Side Setup: The VM hosting machine for Windows 10 was used as the client, as shown in Fig. 17. Client-to-server baseline measurements were taken at iperf and used gigabit connection at rate of 1000 Mbps and a ping time of approximately 1 ms observed over an intranet emulation environment so globally port forward shows 200 ms once delivering outbound to internet. Google Chrome was loaded with the client. Other browsers, such as Firefox Nightly and Safari, also have experimental H3 support but not supported by default as an experimental feature.

Network Impairments: Linux emulation tool NetEm [50] was used to monitor various network parameters, which were crucial in benchmarking the protocols under different conditions. Impairment rules were applied in this analysis to outgoing packets on the network interface of the server. As shown in Fig. 2, both packet loss and delay were considered.

Web Content Served: The website contains CSS, JavaScript, text, and pictures required requests and taking time for serving to client once requested URL is served to client. In this section we provide an experimental evaluation based on HTTP and QUIC. Recently a core stack has been introduced by https://www.litespeedtech.com for faster access as shown in Fig. 18 for which the QUIC streams are beneficial as compared with data streams encapsulated with Lightspeed introduced their HTTP server.

The above proposed QUIC based monitoring using QUIC—HTTP 3.0 probe as virtual engine service has been implemented in beta web application—https://Propsicle.in and resulted in a very good response. The Cloudflare based web proxy was used for testing the QUIC and surprisingly Fig. 19 shows enhanced

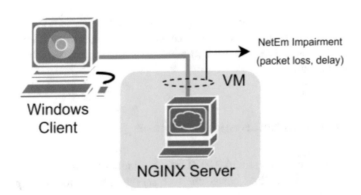

Fig. 17 Network setup for experiment [22]

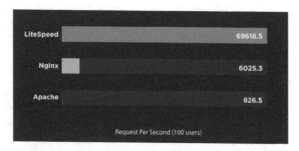

Fig. 18 Core stack data stream-LightSpeed server based on lightspeedtech.com

performance, under 1 s. The timeline for speed index varied due to several restrictions based on internet and adopting earlier version. QUIC.CLOUD is still in development and proposed DNS + Cache, faster response Low Quality Image Placeholder (LQIP), Critical CSS (CCSS), Image Optimization, QUIC CDN can be beneficial once operational over internet. SSL from Let's Encrypt authority also issue free SSL to websites based on QUIC.CLOUD so futuristic prospects and measures can be beneficial. Figures 19 and 20 show insights after implementation and it is observed that there is significant improvement in performance observed instead of normal without QUIC enable applications.

Figure 21 shows drastically decreased URL page load time when data stream backend is with QUIC.

Figure 22 shows that core web vitas for digital marketing. The QUIC protocol is beneficial for web delivering. For search engine algorithms it can be beneficial when web content delivered to clients at faster rates. Resulting visitor drop sessions reduce when delay removed and observing faster experience.

Figure 23 shows load event that is 8 ms and processing for DNS is 91 ms but it's behind cloudflare not with QUIC.CLOUD completely due to beta version of this is in underdevelopment.

On Fig. 24 it shows HTTP3 check Response header showing version 1.1 that is proxies based on CloudFlare inside NGINX based upstream for global. It raised a limitation HTTP doesn't understand by browsers as well servers who adopt HTTP3 QUIC data stream over web delivering in present global internet with https://http3check.net/ as shown in first line in Fig. 24. When beta version will be available finally to public, this can be improved and checked with QUIC support by server headers. Some services for compression available like brotli compression,

🕐 **Critical Loading Times**

Using Cloudflare enhanced performance

	Mobile (3G)	Desktop (Cable)		Mobile (3G)	Desktop (Cable)
Time to First Byte	1.1 s	0.2 s	First Interactive	No data	2.7 s
First Contentful Paint	3.8 s	1 s	Speed Index	9.1 s	3.4 s
First Meaningful Paint	3.8 s	1 s	Total Load time	18.6 s	5.3 s

Fig. 19 Cloudflare based load time from globally inside portal to test load timings

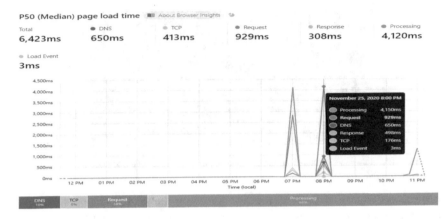

Fig. 20 Browser insights after implementation QUIC.CLOUD with simple HTTP traffic test

Fig. 21 URL report when implementation done with Lightspeed tech based QUIC

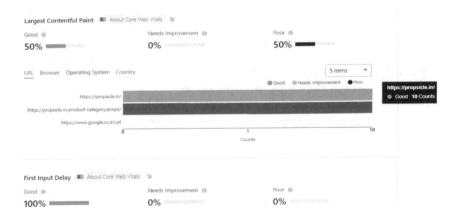

Fig. 22 Core web vitas for digital marketing

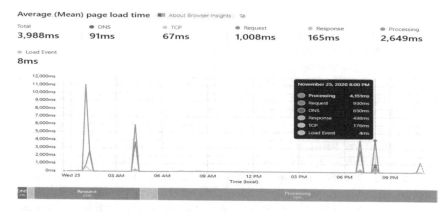

Fig. 23 Average page load time observation when TCP response is 67 ms without QUIC. CLOUD

HTTP Header

```
[HTTP/1.1 200 OK]
Date: Wed, 25 Nov 2020 18:43:59 GMT
Content-Type: text/html; charset=UTF-8
Transfer-Encoding: chunked
Connection: keep-alive
Set-Cookie: __cfduid=d8c6dff259f5bbba71b933fa85296e4301606329839; expires=Fri, 25-Dec-20 18:43:59
GMT; path=/; domain=.propsicle.in; HttpOnly; SameSite=Lax; Secure
Vary: Accept-Encoding,Cookie
Last-Modified: Wed, 25 Nov 2020 18:35:33 GMT
Cache-Control: max-age=3093, public, public
Expires: Wed, 25 Nov 2020 19:35:33 GMT
Strict-Transport-Security: max-age=0; includeSubDomains; preload
Referrer-Policy: no-referrer-when-downgrade
X-Powered-By: W3 Total Cache/0.13.3
Pragma: public
CF-Cache-Status: DYNAMIC
cf-request-id: 06a251485700000ccd30124000000001
Expect-CT: max-age=604800, report-uri="https://report-uri.cloudflare.com/cdn-cgi/beacon/expect-
ct"
Report-To: {"endpoints":[{"url":"https:\/\/a.nel.cloudflare.com\/report?
s=VBRKD2f4LOpeETQhrLJTvVIEqyUBZzEVCk5okqQowXgyM1suLdg0jjv9333ONsK12J00XBzrYIEDM9LnnUupPKIJZ7rl16Q
KR%2F1OSifqFzCZmIPTlVoUcjM%3D"}],"group":"cf-nel","max_age":604800}
NEL: {"report_to":"cf-nel","max_age":604800}
Server: cloudflare
CF-RAY: 5f7d84ba29cb0ccd-EWR
Content-Encoding: gzip
```

Fig. 24 HTTP3 check response header

Gzip deflate for HTTP earlier versions can play only compression for data but not for reducing timeline latency in TCP based HTTP. Only web proxies approach is presently applied.

Due to legacy version is adopted by browsers and maintaining a global equality for relating to delivering web pages, QUIC Protocol is in development phase and only some tuning at client browser only adopts QUIC protocol based on above approach this also implies on simulation only and takes time for globally approach to adopt after several challenges and forecast the drawbacks with overcoming. The new approach that is coming up is software-based devices instead of a dedicated hardware device as it is difficult to install hardware in the network and easy to inject QUIC enabled monitoring probes. Software packages that function much like standard probes are the latest probes. As such, plugging in some dedicated hardware does not require hosting one. Hosts can mount the probes on their own infrastructure bits instead,—e.g. virtual machines, home routers, servers, and so on. RIPE Atlas probes are now available as software, providing a new way for potential hosts to help create the network of RIPE Atlas. Although not a substitute for their hardware counterparts, by taking RIPE Atlas to new and previously difficult-to-reach areas, software probes can boost coverage [51]. It's not an effortless task to implement in operational network where major internet exchanges and multiple Internet Service Providers involved. RIPE Atlas map is implemented with dedicated devices performing tasks but as our proposed probe using QUIC Protocol based. Not much work has been done in this field, the proposed work can be beneficial for further study with QUIC.

7 Conclusions

In order to have faster network response the HTTPs and HTTP can be implemented using QUIC instead of ICMP. This will result in better information technology infrastructure and provide scope for protocol extensibility. In order to adopt new approach to handle data stream hardware and software support is required. It is possible to customise Linux based devices to support QUIC but at present firmware support is not available. The relay based on intermediate border gateway autonomous systems name internet peer nodes also need to be allowed over internet exchanges so that this protocol can plays and act as game changing roles in various field once implemented on internet globally. The proposed approach can be called "QU-ICMP" as ICMP with Quick UDP Internet Connection base is used for services monitoring. For maintaining the roles of probes the approach can also beneficial. For faster wireless, QUIC can be beneficial for network discovery when it comes with 5G and intranet based local networks, easy monitoring can be performed. The peer-to-peer technologies such as VSAT, radio frequencies mobile telecom (UhRF, Vrf, etc.) can also use QUIC for faster communication for HTTP delivering as well navigation systems for real-time monitoring from probes. From the study it can be concluded that the proposed approach is beneficial for network and information technology infrastructure related technologies onshore as well offshore in collocated presence over different IP addresses over public, private and hybrid cloud architectures and monitoring.

References

1. Kumar, P., Dezfouli, B.: Implementation and analysis of QUIC for MQTT. Comput. Netw. **150**, 28–45 (2019)
2. Yosofie, M., Jaeger, B.: Recent progress on the QUIC protocol, seminar IITM WS 18/19. Netw. Architectures Serv. (2019). https://doi.org/10.2313/NET-2019-06-1_16
3. Henderson, T., Floyd, S., Gurtov, A., Nishida, Y.: RFC 3782: The New Reno Modification to TCP's Fast Recovery Algorithm (2012). https://tools.ietf.org/html/rfc3782
4. Xu, L., Harfoush, K., Rhee, I.: Binary increase congestion control (BIC) for fast long-distance networks. In: Proceedings of the Twenty-Third Annual Joint Conference of the IEEE Computer and Communications Societies (INFOCOM), vol. 4, pp. 2514–2524 (2004)
5. Freier, A., Karlton, P., Kocher, P.: RFC 6101: The Secure Sockets Layer (SSL) Protocol Version 3.0 [Online] (2011). Available http://tools.ietf.org/html/rfc6101
6. Atzori, L., Iera, A., Morabito, G.: The internet of things: a survey. J. Comput. Netw. (Elsevier) **54**(15), 2787–2805 (2010)
7. Radhakrishnan, S., Cheng, Y., Chu, J., Jain, A., Raghavan, B.: TCP fast open. In: Proceedings of the Seventh Conference on Emerging Networking Experiments and Technologies, p. 21 (2011)
8. Langley, A., Riddoch, A., Wilk, A., Vicente, A., Krasic, C., Zhang, D., Yang, F., Kouranov, F., Swett, I., Iyengar, J., et al.: The QUIC transport protocol: design and internet-scale deployment. In: Proceedings of the Conference of the ACM Special Interest Group on Data Communication, ser, SIGCOMM '17, pp. 183–196. ACM, New York, NY, USA (2017) [Online]. Available http://doi.acm.org/10.1145/3098822.3098842
9. Dezfouli, B., Radi, M., Chipara, O.: REWIMO: a real-time and reliable low-power wireless mobile network. ACM Trans. Sens. Netw. (TOSN) **13**(3), 17 (2017)
10. Saini, S., Fehnker, A.: Evaluating the stream control transmission protocol using uppaal (2017). arXiv:1703.06568
11. Chimkode, S.V.: Design of an FPGA based embedded system for protecting the server from SYN flood attack, Goa University. Ph.D. thesis (2017)
12. Ahmed, Z., Mahbub, M., Soheli, S.J.: Defense against SYN flood attack using LPTR-PSO: a three phased scheduling approach. Int. J. Adv. Comput. Sci. Appl. **8**(9), 433–441 (2017)
13. LakshmiNadh, D.N., Rao, S.N., Rani, R., Analysis of TCP issues in internet of things. Int. J. Pure Appl. Math. **118**(14) (2018)
14. Bziuk, W., Phung, C.V., Dizdarević, J., Jukan, A.: On http performance in iot applications: an analysis of latency and throughput. In: Proceedings of the 41st International Convention on Information and Communication Technology, Electronics and Microelectronics (MIPRO), pp. 0350–0355. IEEE (2018)
15. Scharf, M., Kiesel, S.: Head-of-line blocking in TCP and SCTP: analysis and measurements. In: Proceedings of the GLOBECOM, vol. 6, pp. 1–5 (2006)
16. Stewart, R.: RFC 4960: Stream Control Transport Protocol [Online] (2007). Available https://tools.ietf.org/html/rfc4960
17. Dierks, T., Rescorla, E.: RFC 5246: The Transport Layer Security (TLS) Protocol [Online] (2008). Available https://tools.ietf.org/html/rfc5246
18. Rescorla, E., Modadugu, N.: RFC 6347: Datagram Transport Layer Security Version 1.2 [Online] (2012). Available https://tools.ietf.org/html/rfc6347
19. Eric Rescorla, H.T.: RFC: 6347 The Datagram Transport Layer Security (DTLS) Connection Identifier [Online] (2017a). Available https://tools.ietf.org/html/draft-ietf-tls-dtls-connection-id-00#page-3
20. Eric Rescorla, H.T.: RFC 6347: Datagram Transport Layer Security Version 1.2 [Online] (2017b). Available https://tools.ietf.org/html/rfc6347
21. Hamilton, R., Iyengar, J., Swett, I., Wilk, A.: QUIC: A UDP-Based Secure and Reliable Transport for HTTP/2. IETF, draft-hamilton-early-deployment-quic-00 (2017)

22. Saif, D., Lung, C.-H., Matrawy, A.: An early benchmark of quality of experience between HTTP/2 and HTTP/3 using lighthouse. arXiv:2004.01978v3 [cs.NI] 11 Oct 2020
23. Viernickel, T., Froemmgen, A., Rizk, A., Koldehofe, B., Steinmetz, R.: Multipath QUIC: a deployable multipath transport protocol. In: 2018 IEEE International Conference on Communications (ICC), Kansas City, MO, pp. 1–7 (2018). https://doi.org/10.1109/icc.2018.8422951
24. Rosenberg, J.: UDP and TCP as the New Waist of the Internet Hourglass. Work in Progress (2008)
25. Cui, Y., Li, T., Liu, C., Wang, X., Kühlewind, M.: Innovating transport with QUIC: design approaches and research challenges. IEEE Internet Comput. **21**(2), 72–76 (2017)
26. Ledkov, D.J., Lallement, J.-B., et al.: Last visited 14 February 2019 [Online]. Available https://wiki.ubuntu.com/CosmicCuttlefish/ReleaseNotes
27. Joseph, A., Li, T., He, Z., Cui, Y., Zhang, L.: A comparison between SCTP and QUIC. IETF, Internet-Draft draft-joseph-quiccomparison-quic-sctp-00, March 2018
28. Paasch, C., Bonaventure, O.: Multipath TCP. Queue, vol. 12, no. 2, pp. 40:40–40:51 (2014)
29. Gont, F., Yourtchenko, A.: On the Implementation of the TCP Urgent Mechanism. RFC 6093
30. Frommgen, A., Erbshäußer, T., Buchmann, A., Zimmermann, T., Wehrle, K.: ReMP TCP: Low Latency Multipath TCP, pp. 1–7. IEEE ICC (2016)
31. Iyengar, J., Thomson, M.: QUIC: A UDP-Based Multiplexed and Secure Transport. IETF, Internet-Draft draft-ietf-quic-transport-14
32. Coninck, Q.D., Bonaventure, O.: Multipath QUIC: Design and Evaluation. Conext'17 (2017)
33. Rezende, P., Kianpisheh, S., Glitho, R., Madeira, E.: An SDN-based framework for routing multi-streams transport traffic over multipath networks. In: ICC 2019–2019 IEEE International Conference on Communications (ICC), Shanghai, China, pp. 1–6 (2019). https://doi.org/10.1109/icc.2019.8762061
34. Viernickel, T., Froemmgen, A., Rizk, A., Koldehofe, B., Steinmetz, R.: Multipath QUIC: A Deployable Multipath Transport Protocol, pp. 1–7. IEEE ICC (2018)
35. Mogensen, R.S., Markmoller, C., Madsen, T.K., Kolding, T., Pocovi, G., Lauridsen, M.: Selective redundant MP-QUIC for 5G mission critical wireless applications. In: IEEE 89th Vehicular Technology Conference (VTC2019-Spring), Kuala Lumpur, Malaysia, pp. 1–5 (2019). https://doi.org/10.1109/VTCSpring.2019.8746482
36. Network Working Group.: IETF, January 1997. https://tools.ietf.org/html/rfc2068
37. Internet Engineering Task Force. https://tools.ietf.org/html/rfc7231#section-6.5.4
38. IETF, Working Group, February 2000. https://tools.ietf.org/html/rfc2775
39. IETF, September 2000, Statement. https://tools.ietf.org/html/rfc2929
40. IETF, November 1996. https://tools.ietf.org/html/rfc2045
41. Section 8.1.3, Reference Point 1 by IETF (RFC 2068)
42. Probe Analysis based on https://www.w3.org/Protocols/HTTP-NG/http-prob.html
43. IETF QUIC Working Group. https://quicwg.org/base-drafts/draft-ietf-quic-http.html
44. IP SWITCH Tool Using for Educational Purpose to Show Local Host HTTP Monitoring
45. IETF J.: Postel RFC Statement on September 1981. https://tools.ietf.org/html/rfc792
46. Source: https://en.wikipedia.org/wiki/Internet_Control_Message_Protocol
47. Source: https://blog.cloudflare.com/introducing-0-rtt/
48. IETF M.: Turner, T.S. (ed.). https://tools.ietf.org/id/draft-ietf-quic-tls-25.html
49. Alessandro Ghedini statement on https://blog.cloudflare.com/the-road-to-quic/
50. Hemminger, S.: Network emulation with NetEm. In: Proceedings of Linux Conference Australia, Canberra, Australia, April 2005 (2005)
51. Viviers, G.: Conference with MANRS at INNOG3 based on Internet Society (MANRS is an Internet Society-supported activity) from RIPE NCC share a brief introduction of probes introduced RIPE Atlas dedicated probes at https://www.innog.net/wp-content/uploads/2020/08/Presentation-2_-RIPE-Atlas.pdf

External Threat Detection in Smart Sensor Networks Using Machine Learning Approach

Oliva Debnath, Himadri Nath Saha, and Arijit Ghosal

Abstract Smart Sensor Networks are receiving a lot of interest for research and at the same time viable applications are beginning to materialize. The biggest challenge of these smart sensor networks is the external threat detection which is a rising concern in the field of smart sensor network in every domain. Smart sensors are always connected with the Internet. Hence these smart sensors are facing huge possibilities to get exposed to different anomalies and attacks which are altogether termed as external threats. These external threats result in incidents like crash of the system. So, external threat detection for Smart Sensor Networks is an important area of research. Denial of Service, Malicious Control, Data Type Probing and Spying are exampling some anomalies and attacks. Machine Learning concept can play an important role to predict different kinds of external threats by making the whole prediction task fully automated. A supervised machine learning approach can be applied to predict the external threats based on previous data history. Several parameters like Source Address, Source Type and Source Location etc. will be able to play the role of feature set for the supervised machine learning approach to predict different kinds of external threats. Support Vector Machine (SVM), Logistic Regression (LR), Random Forest (RF), Artificial Neural Network (ANN) and Decision Tree (DT) are good examples supervised machine learning approaches which can be used for this task. The system proposed here receives a good accuracy which is shown in the experimental result.

Keywords Smart sensor networks · External threats · Machine learning (ML) · Autonomous adaptive detection mechanism · Cyber security

O. Debnath
Department of Computer Science and Engineering, Institute of Engineering and Management, Kolkata 700 091, West Bengal, India

H. N. Saha (✉)
Department of Computer Science, Surendranath Evening College, Kolkata 700 009, West Bengal, India

A. Ghosal
Department of Information Technology, St. Thomas College of Engineering and Technology, Kolkata 700 023, West Bengal, India

© The Author(s), under exclusive license to Springer Nature Switzerland AG 2022
U. Singh et al. (eds.), *Smart Sensor Networks*, Studies in Big Data 92,
https://doi.org/10.1007/978-3-030-77214-7_7

1   Introduction

The Smart sensor networks have been developed rapidly in the recent years. The concept of smart devices is a system of interrelated computing devices, which are ingrained with electronics, sensors, software, network connectivity which when comes together allows these devices to exchange data. The smart sensor networks continue to extend its barriers by establishing a connection between the cyber world and the physical world.

Smart sensor networks have been implemented in various devices and applications and are the next major thing in the domain of emerging technologies. The application of smart sensor networks is highly favoured today and their usage can be found in every field like Smart cities, Smart vehicles, Smart Retail, Smart Healthcare, Smart Agriculture and the list goes on. These smart systems offer a plethora of benefits when it comes to managing the urban infrastructure of today. The Smart sensor networks therefore, are contributing to a great deal to the common people and making their lives easier.

The stride of connecting physical devices around us to the Internet [1] is on a rapid increase today. According to a recent report of the Gartner, it is found that the consumer segment is the largest user of connected things with 12.86 billion units in 2020 representing 63.04% of the estimated total number of applications in use. It forecasts that 20.4 billion connected things will be in use by 2020. Therefore, the application of smart sensors are increasing all over the world. Along with the vast use of connected devices, comes the issues of security of these systems and their privacy. However, the tasks with huge volumes of data and high computational complexities are prone to security breaches due to constraints in resources. The huge amounts of data generated [2] and exchanged by the IoT devices put a lot of strain on the IoT Networks. Security issues regarding these Smart Sensor Networks also cause several issues in operation with high profile attacks, mass exploitation of devices, and eye-catching headlines about "exotic" device hacking [3]. One of the major issues with these networks is the varied nature of its deployments. This heterogeneity in the smart sensors gives rise to many challenges particularly when it comes to security and privacy. As the number of devices grow, the amount of data will increase. The smart sensor networks are therefore growing vulnerable to the external attacks.

The first step to providing this security is the detection of these external threats. Without a trustworthy environment, the high demand of these devices may be lost and the devices may lose their potential too. The Internet of Things is going to be the most important source of new data that is going to make more intelligent applications of smart sensors. Security is highly needed in all the Smart sensor networks that are already deployed or will be deployed in future. The use of smart sensors in our day-to-day activities are increasing at a humongous rate. Therefore, the Security concerns are in demand right now.

Machine Learning (ML) has attracted a noticeable crowd in the field of research. It has the perspective to change the direction of cyber security as attack detection in smart environments providing us with promising results. Algorithms have been developed through Machine Learning for providing solutions against these cyber-attacks. It gives us a new approach to protect our systems and prevent them from crashing.

Smart Sensor Networks usually look after environments, which are dynamic [4]. So, it is important to design these networks in such way that they can adapt with the new environment efficiently. An autonomous adaptive detection mechanism is implemented here to overcome these attacks. Each of these attacks has been further classified in order to create a more granular prediction model. Supervised Machine Learning Algorithms like Random Forest (RF), K Nearest Neighbour (KNN), Logistic Regression (LR) have been used to predict some of the major category of attacks which include Denial of Service, Probe Attacks (Datatype Probing), Privilege Escalation Attacks and Remote Access Attacks. A robust model has been proposed in this book chapter which caters to a huge array of external threats and their detection. The model was observed to attain an accuracy of 99.321% giving the best results when the attacks were classified by a Random Forest Classifier.

Section 2 consists of a literature survey on the related works in the field of security of these Wireless Smart Sensor Networks (WSNs) mentioning their results and their pros and cons. The External Threats occurring in the devices using smart sensors have been elaborated. In Sect. 3 our proposed model has been elaborated. It also depicts the machine learning algorithms used here for the detection of external threats. Section 4 deals with the methodology used in our proposed model. In Sects. 5 and 6, a comparative analysis with respect to other Machine learning Algorithms followed by the Performance Analysis has been put forth. Finally, Sects. 7 and 8 caters to the future research directions and a short conclusion of our book chapter.

2 Background and Related Works

The smart sensor networks are performing tasks which are generating huge volumes of data. Due to constraint in the resources, the high computational complexities of these networks are prone to compromising the security of these networks. The data generated and exchanged by the IoT devices is huge which puts a lot of strain on the IoT Networks.

Oscar Novo [5] proposed an architecture to secure the IoT devices by using Block Chain Technology for attaining a Decentralised Access on IoT Networks. It is a fully distributed access control system for smart sensor devices based on blockchain technology. But blockchain technology uses a lot of energy and cannot be implemented in a huge distributed system. Hassija et al. [6] have performed a survey on IoT security. For survey they have considered different applications of IoT. In this work they have assessed different security related challenges in IoT applications along with different sources of threats.

Canedo and Skjellum [7] have implemented neural networks to develop a machine learning model that can measure the validity of the information received from IoT devices. But the data required to feed the network is much larger than what is required by supervised learning algorithms. The data is also required to be labelled which is another issue in implementing machine learning models working with neural networks. Zhang et al. [1] has proposed a framework to detect and solve the cross layer wireless attacks based on Bayesian Theorem. But constructing data from Bayesian networks are not universally accepted as a method of data collection.

Musa G. Samaila, Joao B.F. Sequeiros [3] along with others had proposed a framework and a guide for Secure Design and Development [3] of devices and applications. They have recommended specific Light weight Cryptographic Algorithms (LWCAs) for implementing them on both software and hardware. The limitations here mainly include the lack of enough LWCAs for use in the LWCAR component of the proposed framework. There has not been any evaluation of these algorithms and no algorithms have been included in the framework for the same. Machine Learning includes Reinforcement learning which allows machines to automatically find out the ideal behaviour within a particular context to maximise its performance. It is about the interaction between two elements—the environment and the learning agent. If the machine makes the correct prediction, it gets a reward or else it gets a penalty. In this chapter we have experimented with three very popular Supervised Machine Learning algorithms for the detection of attacks as a solution to the security requirements in the smart sensor networks which are Random Forest (RF), K-Nearest Neighbour (K-NN), and Logistic Regression (LR).

Aghajan et al. [8] have employed wireless sensor network with multiple sensors and ability to recognize events. Aim of their work was to assist vulnerable people by reducing chances of occurrence of accidents in the home. Distributed vision-based analysis bundled with AI-based algorithms has been adopted in their work for assisted living.

Arth et al. [9] have worked with remote traffic surveillance along with surveillance of distant places under harsh environmental situations. They have used embedded platform for this purpose. The center of attention of their work was to detection and tracking of vehicles from the captured images. Image authentication techniques have been handled by Albanesi et al. [4] in their work. Aim of their work was to design safe, strong and efficient authentication algorithms.

They have proposed taxonomy for this task which considers three features. Capability of localizing manipulated areas, level of integrity verification and the approach to the generation of the authenticator has been considered as the features in their work.

Security issues regarding these Smart Sensor Networks also cause several issues in operation with high profile attacks, mass exploitation of devices, and eye-catching headlines about "exotic" device hacking [10]. The heterogeneity in the smart sensors and the nature of deployment, gives rise to many challenges particularly when it comes to security and privacy. As the numbers of devices grow, the amount of data keeps increasing. Therefore, the smart sensor networks are growing highly vulnerable to the external threats. Wireless Transmissions are extremely vulnerable and they are open to be attacked by different external threats which include Denial of Service (DoS), Malicious Code Injection, Jamming and others. Although measures have been taken to overcome these attacks and several solutions have been proposed, they are all streamlined to some particular behaviour of these attackers and particular wireless technologies. Solutions have also been given involving the use of Cryptography but they are not practically applicable since the smart sensor environment is highly resource constrained.

Atrey et al. [11] have suggested a healthy hierarchical scheme for the purpose of video authentication. Their work was fully dependent on cryptographic secret sharing. Cryptographic secret sharing was used to prevent a video from temporal jittering as well as spatial cropping. The authors claimed that their algorithm provides a trade-off between heftiness and security by having constructible inputs.

There are a lot of solutions existing for threat detection in the smart sensor nodes. But these existing solutions are not sufficient owing to the unique character of the nodes. This is because of several challenges faced by the smart devices including the resource constraints, the humongous amount of data generated by them, heterogeneity and their dynamic behaviour. Therefore, Machine Learning (ML) [12] techniques are a need of the hour which is capable of providing security and intelligence in the smart sensor networks. One of the popular approaches for a machine learning algorithm is supervised approach where the machine or system is trained with a labelled or an unlabelled dataset to produce a model. New input data is introduced into system through the machine learning algorithm for testing the model and it makes predictions based on the model. Prediction's accuracy is checked to see if it is acceptable and then it is deployed into the Smart sensors or devices. Machine Learning has three types:

(i)   Supervised Learning
(ii) Unsupervised Learning and
(iii) Reinforcement Learning.

Aaraj et al. [13] have proposed a system to improve the security of general computing systems. Goal of their work was to analyse and plan a hardware and software trusted platform. In their work, they have evaluated energy and carrying out time overheads for a SW-TPM. To achieve their goal, they have always considered to ensure that computational and energy constraints for SW-TPMs are within acceptable range.

In Supervised Learning [14], authors have an input variable 'x' and an output variable 'y' and we use an algorithm to learn the mapping function from the input to the output i.e. $y = f(x)$. Once the algorithm is trained, it can be used to predict the correct output of a never seen input. Some of the popular Supervised Learning Algorithms are: Linear Regression, Logistic Regression, Decision Trees, Random Forest, K Nearest Neighbour, and Support Vector Machines.

There can be a variety of classes of attacks on these smart sensor networks. The model implemented here, focuses on the major classes of attacks imposing threats on the IoT system. The main objective of this book chapter is to study about these attack types in detail for an in-depth understanding. We have four major attack types:

(A) Denial of Service (DoS)
(B) Probe Attacks
(C) Privilege Escalation Attacks or Privilege Attacks and
(D) Remote Access Attacks.

(A) Denial of Service Attacks:

In Denial of Service attacks, the target server is flooded by the attacker with a large number of unwanted requests. This makes the target server incapable of allowing any more requests into the system. Thus, the requests of the genuine users get blocked in the way leading to disruption of the services. Jamming is one of the popular attacks under Denial of Service attacks. A malicious code can intentionally interfere in the smart sensor networks blocking some legitimate communication or data. Game theory [14] has been proposed as a well-suited mathematical framework for providing effective anti-jamming [14] techniques. But there has not been any implementation of the framework. Replay attacks [14] aim at creating congestion in the network by eavesdropping and re-transmitting several copies of the same packet. Timestamps have been suggested as a possible solution of these attacks but there can be a major disadvantage to this too. If the attacker knows how to generate an allowable timestamp, then timestamps cannot be used to ensure security for Replay attacks. Table 1 below shows the various possible Denials of service attacks for smart sensor network.

(B) Probe Attacks:

In case of probe attacks in smart sensor network, the attacker illegally investigates into the network used by the smart devices and gains valuable information and data about the activities occurring in the network. The attacker then uses the data to exploit the network and crash the systems. Probe attack is also termed as probe-response attack. Probe attack results in a kind of intrusion to the system. Supervised machine learning approach may be used to identify this probe attacks as the attacked system changes its behaviour compared to normal system. Below mentioned Table 2 reflects the various kinds of probe attacks that these smart sensor networks can be exposed to:

(C) Privilege Escalation Attacks:

In case of Privilege Escalation Attacks in smart sensor network, the attacker, who is termed as malicious user, gains unauthorized elevated access privilege to access the system within the security parameters, or within the sensitive systems which are usually protected from an application or a user. The attacker gains the access right

Table 1 Various possible Denial of service attacks on the smart sensor networks

DoS attacks	Attack description
apache2	The attacker here sends a request to an Apache Server to retrieve the URL content (Uniform Resource Locator) in a large number of overlapping bytes. The server therefore runs out of usable memory resulting in a denial-of-service condition
back	The attacker here, requests for a URL from a web server. The URL requested by the attacker will be one which has many backslashes
land	The hacker here sends a UDP (User Datagram Protocol) packet having the same source and destination address to the remote host. The host computer thus gets locked up causing a denial-of-service condition
neptune	The attacker floods the SYN on ports in this case. The receiving host uses up its resources waiting for the session to start. In this process, the receiving host becomes unavailable for the legitimate traffic causing a denial of the services
smurt	A smurt attack is a Distributed Denial of Service attack (DDoS) in which the echo reply of the ICMP (Internet Control Message Protocol) is flooded. As a result, the target system becomes unable to operate because it slows down to a point that it becomes vulnerable
mailbomb	A mailbomb is a kind of attack where massive amounts of emails are sent by the cybercriminal to a specific user causing their system to slow down. This results in denial of services by that system
pod	'pod'stands for Ping of Death. It is a type of dos attack in which the attacker pings the hosts system with malformed packets causing the system to reboot or crash
udpstorm	It is a dos attack where a large number of packets of the UDP (User Datagram Protocol) are sent to the targeted server aiming to crash the system
teardrop	In a teardrop attack, the hacker sends broken and disorganized IP (Internet Protocol) fragments with overlapping bytes and oversized payloads to the victim's machine in order to crash the Operating system
worm	A worm is a malware program that can replicate itself and spread to other systems with the help of a network. The security of the network is compromised. The attacker uses the target machine as a host to scan and spread the malicious software or the malware program into other systems

of administrator from a simple user in an escalated way. Hence this attacked is termed as *Privilege Escalation Attack*. This privilege escalation is of two types:

(i) Horizontal privilege escalation and
(ii) Vertical privilege escalation

In horizontal privilege escalation, attacker gains the access of functionality and data of another user in same level. In vertical escalation, attacker gains the access of functionality of administrator or any power user. Vertical escalation is also termed as privilege elevation.

The attacker can then employ this unauthorizedly gained privilege to access confidential data and finally damages the system. Table 3 contains some of the Privilege Escalation Attacks occurring in the smart devices.

Table 2 Various possible Probe attacks on the smart sensor networks

Probe attacks	Attack description
satan	'satan' is a kind of probe attack where the hacker scans the network unlawfully for some well-known weaknesses in the topology design, or some well-known ports and then uses them to damage the system
ipsweep	This is a kind of attack which involves pinging of multiple hosts to reveal the IP addresses of the target node. The attacker sends ICMP (Internet Control Message Protocol) echo requests to multiple hosts. If the target host replies, the reply contains the IP address which gets revealed to the attacker causing a probing attack
portsweep	In a portsweep attack, the attacker sends client requests to a range of server port addresses on a host. It scans through the ports to discover services on a host, and then exploits the channels and the communication
nmap	It is a means of network mapping which can be used by a cyber-criminal to identify all the devices connected in the network and gather valuable information about the ongoing services and the Operating system
saint	The IoT applications are hacked by saint attacks where an attacker gains important information about the network by site scanning
mscan	This is a kind of probing where the hacker exhaustively scans the IP address to discover the vulnerabilities in the smart sensor networks

(D) Remote Access Attacks:

These are cyber-attacks where the cyber-criminal tries to get remote access of the system and finally gets a hold of vulnerable points in the security software of the network. Therefore, the cyber-criminal gains access to the device or the system can inject malicious code into the program causing a serious threat to the security of the system [15]. Various types of Access attacks that can occur in the smart sensor networks are demonstrated in Table 4.

Mohammad Abu Alsheikh, Shaowei Lin, Dusit Niyato Hwee-Pink [16] elaborated the issues occurring in the wireless sensor networks. They suggested models using Machine Learning which could be used to address the issued faced by the wireless sensor networks. But the methods were not sufficient because it is required to keep in mind that the nodes need to be charged at regular intervals. They are very prone to the hackers which make the communication speed very low disrupting the functionality of the networks. In Unsupervised Learning, there is only the input variable x but no corresponding output variable. Unsupervised ML models the unsupervised structures in the data in order to gather more information about the data. Unsupervised Learning algorithms detect a pattern based on innate characteristics [16] of the input data e.g.: Clustering – Here, similar data instances are grouped together in order to identify clusters of data. Some of the popular unsupervised learning algorithms are: Apriori algorithm, Hierarchical Clustering and K-Means Clustering.

Table 3 Various possible privilege escalation attacks on the smart sensor networks

Privilege attacks	Attack description
buffer_overflow	This is a kind of privilege attack where an attacker uses the buffer overflow issues to overwrite the internal memory of an application. The execution path of the program changes due when a buffer_overflow attack occurs resulting in damage of the files and exposure of private information of the program to the attacker
rootkit	'rootkit' is a software which is used by the attacker to have unauthorized privileged access to the target network and the restricted areas of the software
load_module	The attacker here gains access of the root shell and resets the IFS (Internal Field separator) of the system
perl	In a perl attack, the hacker creates a root shell which sets the user id to the root. These files now have suid (set user id) permissions and can run with higher privileges. The attacker uses this to run the program with root privileges giving rise to a privilege escalation attack
ps	It is a PowerShell attack where the attacker incorporates into the PowerShell and can change the commands of the program, thereby hacking the system and disrupting the services of the devices
sqlattack	This is an SQL injection attack where the hacker has the capability to insert malicious SQL statements in the program or changed the records in the database. In this way the private data of the user is compromised
xterm	'xterm is a kind of privilege attack where the attacker first gains control of the system and then uses it to change the connection settings. The attacker fills up the host field with garbage values which prevents the sensors from working properly

Restuccia et al. [15] has paid emphasis on the security threats of IoT in their work. For this purpose, they have proposed to use machine learning to protect IoT devices from different security threats. Sicari et al. [17] has also pointed out security aspect in IoT in their survey paper. Da Xu et al. [18] has performed a survey on IoT from the point of view of industrial approach. They have discussed about the current trends of IoT in industry. Zou et al. [19] has reviewed the weak points on the point of view security and threats in case of wireless communication. They have discussed different technical challenges of security for different wireless communications like Bluetooth, Wi-Fi, WiMAX etc. Mpitziopoulos et al. [20] has carried out a survey on the jamming attacks for wireless sensor networks. Jamming attack is considered as most severe security concern for wireless sensor networks. They have discussed different common jamming techniques in the survey.

Granjal et al. [21] has also performed survey on IoT on the point of view of security. They have discussed different protocols which are existing in IoT. They have also highlighted the open research issues on security point of view in IoT. A joint framework-based approach is recommended by Zhou et al. [22] which comprises of both physical and application layer security technologies in order to provide security for wireless multimedia systems. Zhang and Melodia [2] have also

Table 4 Various possible remote access attacks on the Smart sensor networks

Remote access attacks	Attack description
warezmaster	Illegal software is known as warez. In a 'warezmaster' attack, the attacker uploads warez on the FTP server. The system is then exploited providing wrong write permissions disrupting its normal functioning
warezclient	The attacker downloads the illegal software (warez) uploaded previously by the warezmaster. Warezmaster and warezclient are remote to local attacks (R2L) which exploits the vulnerabilities that are present in 'anonymous' FTP's on both Windows and Linux
guess_password	This attack involves the attacker to guess the password of the target system using cryptographic functions thereby hacking the system, gaining access to the private information
imap	It is extremely dangerous because it allows the hacker to gain access to the local user's account and use the vulnerabilities to damage the system
ftp_write	An improperly configured FTP server is exploited by the attacker in which the home directory is not writing protected. The attacker creates a rhost file in an 'anonymous' FTP to obtain local login to the system resulting in an Access Attack
multihop	It is a kind of routing attack where the cyber criminals combine a range of threats set at multiple stages across multiple points of entry (attack vectors). The aim is to infect the network and the system
phf	The cyber-criminal executes arbitrary commands on a machine with a misconfigured web server. Therefore, he can illegally gain access to the system resources and damage the functionality of the system
spy	The hackers here, breaks into the system via vulnerabilities to gain access to important and private information of the network
named	The attacker gains illegal access to the clients account or the device through improper means and can exploit the user's account or the network
sendmail	A malicious spammer can use the send mail attack to send large number of mails anonymously. This can slowdown the system. It is a security vulnerability that can occur in the IoT applications by overflowing of the mail box with spam mails
snmpgetattack	snmp stands for Simple Network Management Protocol. This is an Access attack where the cyber-criminal can hack the system and gain access to the SNMP to monitor the network and disrupt its management
snmpguess	This is an access attack where the he hacker intrudes into the system by guessing the SNMP and can hamper the performance of the network
http_tunnel	The HTTP tunnel is a data driven access attack which exposes the network connections without exposing the internal system to direct attack. The network connections can be altered creating an unnecessary traffic in the network
xclock	The xclock command gets the time from the system clock and then displays and updates it in the form of a digital or analog clock. The hacker uses this command to alter the settings of the time. Therefore, a xclock attack is launched which delays the responses of the Smart devices
xsnoop	The attacker attempts to exploit the data from the internal memory of the program. He gains unauthorized access which allows him to extract data from the cache illegally

worked with security threats in wireless sensor network. They have dealt with hammer and anvil attack in their work. This attack is nothing but a jamming aided attack which is a severe attack in wireless sensor network.

Zhang et al. [23] have worked with Cross-layer Attacks in case of wireless network. In their work they have tried to identify these attacks and then they tried to resolve those attacks. This attack uses jamming attack which is great security concern. Ashton [24] have discussed on the importance of RFID in case of IoT. Singh and Kapoor [25] have conducted a survey on the platform of IoT. They have also studied different IoT hardware items as well as software platforms.

So, what we need is an effective solution which can easily be deployed in the Smart Sensor Networks for the detection of these external attacks. The class of attacks needs to be validated and investigation needs to be done in this matter. Moreover, the effects of these attacks on the smart networks needs to be analysed. If an attacker hacks the system and gains control on the training data, he could use this data to plan his attacks accordingly. Thus, such attacks can be costly to the system. In this book chapter, a system has been implemented where the classes of attacks have been validated and interesting observations have been made from the same. Supervised Machine Learning Algorithms have been used to train our data for attack detection before they are deployed in the smart sensor nodes. The predictive model proposed here has also been analysed to find out the various services that are present in the attack set. Several parameters like duration, protocol type, service, flag, source bytes, destination bytes have been used as labels which play the role of a feature set for our Supervised Machine Learning Approach.

3 Proposed Model

The model (in Fig. 1) that has been proposed can detect external threats occurring in smart sensor networks. It goes through several steps before the model is fitted and evaluated.

Data Collection and Data Pre-processing is a major step when dealing with Machine Learning Algorithms. It is the first step before the data is fed to the model to make predictions. After the data is processed and transformed, classification of each of the kinds of attacks are done. Then comes the step of Data Analysis. It is primarily done to analyze the data that we are working with and gain a better understanding of the data. It is the initial step in determining what insights the data can yield when we feed it to a machine learning model to make predictions. In our case, it detects the external threats that are present in the Smart Sensor Networks. Following this is the step of Feature Engineering. It is a process to transform the raw data in terms of features. After Feature Engineering, it is time to fit the model to train it and then test it to see if it can detect the threats occurring in the network. Our Estimators include: a) Random Forest Classifier b) K-Nearest Neighbor Classifier and c) Logistic Regression Classifier.

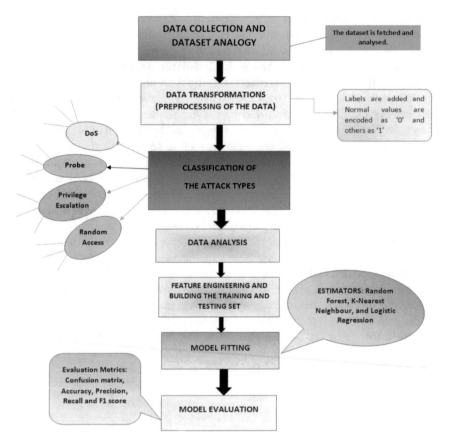

Fig. 1 Steps of the advised approach

(a) **Random Forest (RF)**: It is a versatile algorithm capable of performing both Regression and Classification. It is a kind of an ensemble learning method commonly used in predictive modelling and ML techniques. It compiles the results of multiple decision trees and gives the final results. The first step is to divide the data into random smaller subsets. Every subset may not be distinct, they may overlap too. Then, decision trees are made and the final result is obtained from majority voting. It was chosen in our experiment because it works efficiently on large databases. Therefore, it is scalable. It performs implicit feature selection and has methods for balancing the errors in an unbalanced dataset. It is the most accurate learning algorithm and it works equally well for both classification as well as regression problems.

(b) **K-Nearest Neighbor (K-NN)**: KNN is a simple algorithm that stores all the available cases and classifies the new data or case based on a similarity measure. Here, K = the number of nearest neighbors which are the voting class of the new data or the testing data. Say, if K = 1, then the testing data are given the same data as the closest example in the training set (only 1 nearest neighbor is selected). If K = 3 the labels of the three closest classes are checked and the most common label is assigned to the testing data. In KNN Algorithm, majority voting is done like in RF after selecting 'K' number of nearest neighbor (s) and the final result is computed. It is called a lazy learner because there is no learning phase of the model here. It memorizes the training data, so of the work happens when the prediction is requested.

(c) **Logistic Regression (LR)**: It is a supervised ML algorithm which produces results in a binary format which is used to predict the outcome of a categorical dependent variable [3]. It can perform multi class classification. It gives the probability of the occurrence of an event (in our case, an external threat is the event). The general steps (in order) to implement logistic regression are: Collecting the data, analysing the data, Data wrangling or cleaning of the data (say if we have null values or NAN values), training, testing and accuracy check.

After the models are trained and tested using the above algorithms, their accuracies are checked. So, the final step is for Evaluation of the Model. We have chosen a few Evaluation metrics like Confusion Matrix, Accuracy, Precision, and F1 Score to analyze the performance of our model. The models have shown a high level of accuracy of 99.3% for RF, 98.2 for KNN and 98.3% for LR. Therefore, they can be deployed in the smart sensor networks for the successful detection of threats.

3.1 Algorithm and Process Flow of the Proposed Intrusion Detection Method

The process flow of the proposed Intrusion Detection method is depicted along with the algorithms required for the procedures.

ALGORITHM 1: DATA COLLECTION
Input: NSL KDD dataset
Procedure:
1: import libraries- Numpy, pandas, Matplotlib.pyplot, seaborn, itertools, random
2: import modules from sklearn library – RandomForestClassifier, KneighborsClassifier & LogisticRegression
3: import LabelEncoder, train_test_split,cross_val_score, mean_absolute_error, accuracy_score, Confusion_matrix
4: Fetch the training file (NSL KDD) in csv format

Output: The dataset is fetched in csv format

ALGORITHM 2: DATA TRANSFORMATION

Input: Column labels
Procedure:
1: columns← 'duration', 'protocol_type', 'service', 'flag', 'src_bytes' and so on
2: attack_flag← 0 or 1 based on normal and attack respecticely
3: display attack_flag

Output: attack_flag column is shown with normal values encoded as 0 and any other value as 1

ALGORITHM 3: ATTACK CLASSIFICATION

Input: lists to hold attack classifications
Procedure:
1: list 1: dos_attacks← 'apache2', 'back', 'land', 'neptune', 'mailbomb', 'pod','processtable', 'smurf', 'teardrop', 'udpstorm', 'worm'
2: list 2: probe_attacks← 'ipsweep', 'mscan', 'nmap'
3: list 3: privilege_attacks← 'buffer_overflow', 'loadmodule', 'perl', 'rootkit', 'sqlattack', 'xterm'
4: list 4: access_attacks← 'warezmaster', 'warezclient', 'ftp_write', 'guess_passwd'', 'http_tunnel', 'imap', 'multihop', 'named', 'phf', 'sendmail', 'snmpgetattack', 'snmpguess', 'spy', 'xclock', 'xsnoop'
5: define function "map_attack(attack)"
          Conditions: for dos attacks:
        attack type←1
             for probe attacks:
        attack type←2
             for privilege attacks:
        attack type←3
             for access attacks:
        attack type←4
          for all others:
        attack type←0
   return attack type
6: Store attack_map← attack_type
7: Display 'attack_map'
Output: attack_map column shows the values accordingly by which the nature of attacks which need to be mitigated can be understood

ALGORITHM 4: DATA ANALYSIS
Input: titles to make pie charts
Procedure:
1: define function bake_pies(data_list, labels): for drawing pie charts
2: build subplots: fig,axis←plt.subplots, update the color with new values,
    set the title and labels of the pie charts: title←Flags, location← centerleft,
    return axis
3: build a crosstab to get attack vs protocol to get the series for each protocol icmp,
udp and tcp and build pie charts using bake_pies function using labels icmp, tcp, and
udp
4: build pie charts using bake_pies function for normal_flags and attack_flags
(where attack flag=0 & 1 respectively) using labels normal and attack
5: build pie charts using bake_pies function for normal_services and
attack_services(where attack flag=0 & 1 respectively) using labels normal and attack
6: display the plots using plt.show

Output: Pie charts are generated and displayed indicating a lot of variance for
features - 'service' 'attack_flag', and 'protocol_type'.

ALGORITHM 5: BUILD THE TRAINING MODEL AND MODEL EVALUATION
Input: attack_flag for binary classification and attack_map for multiclassification
Procedure:
1: define the list of models-
    models RandomForestClassifier(), KneighborsClassifier( ) & LogisticRegression(max_iter=250)
2: create a list to store the performance of the models
3: Calculate and display the prediction accuracy of each of the models with
Confusion matrices for binary classification

Output: Training models are run and Accuracy and confusion matrices displayed
for each model.

3.2 Process Flow of the Proposed Methodology

The process flow of the proposed methodology (Fig. 2) involves a few steps in
detecting the external threats attacking the smart sensor networks. These include
fetching the training dataset at first. The next step is to add column labels showing
normal attacks as 0 and all others as 1. This would help in distinguishing the attacks
that needs to be taken care of. In order to get a better visualisation of the features
that are highly varying, pie charts need to be made to understand the network traffic
more clearly. The pie charts show the variation in the features such as which

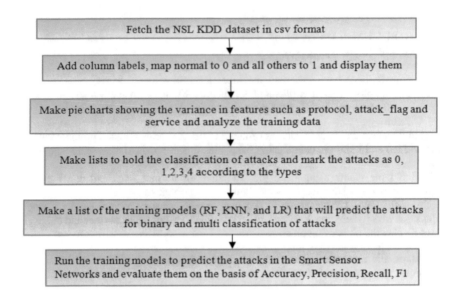

Fig. 2 Process flow

protocol is showing the maximum number of attacks, and the service used in that protocol which is most prone to external threats.

Next, attacks are classified as Denial of Service attacks, probe attacks, access attacks and privilege attacks and normal attacks. To distinguish them, they need to be marked. Here, we have marked as 0, 1, 2, 3, 4 respectively based on the attack types. Now our model is ready to be tested for prediction. Random Forest, KNN, and LR are used to make prediction of the attacks and they are evaluated to see which of them yield the best prediction results i.e., showing the highest accuracy in prediction. For evaluation purpose, few evaluation metrics have been used like Accuracy, Precision, Recall, F1 score for a clear understanding of the differences in each of the predictions, and to also see how much the results deviate in each case. After the evaluation of the model is done, we can conclude that a certain deep learning algorithm works best with this dataset and can be launched to further mitigate attacks occurring in smart sensor networks.

4   Methodology

The experiment was done on a Lenovo ideapad 510 where the operating system was Windows 10 Enterprise 64–bit. The hardware specifications used for running the model are summarized in Table 5.

For Data Analogy, data profiling and feature engineering Numpy module, Pandas module and itertools module were used, for Data Visualization Matplotlib

Table 5 Hardware specifications used for running the model

Elements	Hardware specifications
Processor	Intel (R) core (TM) i5-7200U
CPU	@ 2.50 GHz 2.70 GHz
RAM	8.00 GB
Graphics Card	NVIDIA GeForce 940 MX 2 GB

module, Seaborn module were used and lastly for model fitting and evaluation scikit-learn module and random module were used.

4.1 Data Collection and Dataset Analogy

The NSL KDD dataset was used for Intrusion detection research. The NSL KDD dataset is modifiable, extensible, and reproducible. It reflects the traffic compositions and intrusions. The dataset has 1,25,973 training records and 22,544 testing records each having 41 features such as duration, protocol type, service flag, source bytes, destination bytes, etc. The traffic distribution on NSL KDD in two class and five class are shown in Tables 6 and 7 respectively.

For our experiment, we have taken the dataset in two classes (Normal vs Attack) and five class (Normal, DoS, Probe, Privilege, and Access attacks). We started by fetching the NSL KDD dataset which is available in csv format over Kaggle. A few libraries were imported in the beginning which were numpy, pandas, matplotlib. pyplot, seaborn, itertools and random. Some model imports were done from the scikit-learn library which include Random Forest Classifier, K-Neighbors Classifier and Logistic Regression. Finally, some processing imports were done which were LabelEncoder, train_test_split, cross_val_score, mean_absolute_error, accuracy_score and confusion matrix.

Table 6 Traffic distribution of NSL KDD in two classes

Traffic	Training	Testing
Attack	67343	9711
Normal	58630	12833
Total	125973	22544

Table 7 Traffic distribution of NSL KDD in five classes

Traffic	Training	Testing
Normal	67343	9711
DoS	45927	7458
Probe	11656	2754
Privilege	52	200
Access	995	2421
Total	125973	22544

4.2 Data Transformation

Step 1: Labels were added to the dataset at first because the dataset doesn't include them.

The labels that were added are shown in a snapshot in Fig. 3 followed by the output of that in Fig. 4.

Step 2: Transformations were done around the attack field. The normal values have been encoded as '0' and any other value as '1'. This is used as a classifier for the binary classification of attacks. "attack_flag" was used to denote these values as 0 and 1. The output after the flag was added as the last column is shown in a snapshot in Fig. 5.

4.3 Classification of Attacks

Each of the attack types were classified for a more creating a granular prediction model. Lists were made in the model to hold the attack classifications.

The classification of the four attack types (Denial of service, Probe attacks, privilege escalation attacks and access attacks) are shown in Fig. 6a–d respectively.

```
# add the column labels                    , 'count'
columns = ([ 'duration'                     , 'srv_count'
. 'protocol_type'                           , 'serror_rate'
. 'service'                                 , 'srv_serror_rate'
. 'flag'                                    , 'rerror_rate'
. 'src_bytes'                               , 'srv_rerror_rate'
. 'dst_bytes'                               , 'same_srv_rate'
. 'land'                                    , 'diff_srv_rate'
. 'wrong_fragment'                          , 'srv_diff_host_rate'
. 'urgent'                                  , 'dst_host_count'
. 'hot'                                     , 'dst_host_srv_count'
. 'num_failed_logins'                       , 'dst_host_same_srv_rate'
. 'logged_in'                               , 'dst_host_diff_srv_rate'
. 'num_compromised'                         , 'dst_host_same_src_port_rate'
. 'root_shell'                              , 'dst_host_srv_diff_host_rate'
. 'su_attempted'                            , 'dst_host_serror_rate'
. 'num_root'                                , 'dst_host_srv_serror_rate'
. 'num_file_creations'                      , 'dst_host_rerror_rate'
. 'num_shells'                              , 'dst_host_srv_rerror_rate'
. 'num_access_files'                        , 'attack'
. 'num_outbound_cmds'                       , 'level' ])
. 'is_host_login'
. 'is_guest_login'
```

Fig. 3 Column labels added in the dataset

	duration	protocol_type	service	flag	src_bytes	dst_bytes	land	wrong_fragment	urgent	hot	...	dst_host_same_srv_rate	dst_host_diff_srv_rate
0	0	udp	other	SF	146	0	0	0	0	0	...	0.00	0.60
1	0	tcp	private	S0	0	0	0	0	0	0	...	0.10	0.05
2	0	tcp	http	SF	232	8153	0	0	0	0	...	1.00	0.00
3	0	tcp	http	SF	199	420	0	0	0	0	...	1.00	0.00
4	0	tcp	private	REJ	0	0	0	0	0	0	...	0.07	0.07

5 rows × 43 columns

Fig. 4 Output of the column labels added in the dataset

dst_host_srv_diff_host_rate	dst_host_serror_rate	dst_host_srv_serror_rate	dst_host_rerror_rate	dst_host_srv_rerror_rate	attack	level	attack_flag
0.00	0.00	0.00	0.0	0.00	normal	15	0
0.00	1.00	1.00	0.0	0.00	neptune	19	1
0.04	0.03	0.01	0.0	0.01	normal	21	0
0.00	0.00	0.00	0.0	0.00	normal	21	0
0.00	0.00	0.00	1.0	1.00	neptune	21	1

Fig. 5 The output with the last column showing the values of attack_flag

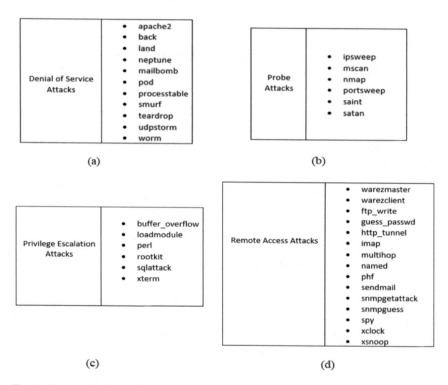

(a)

(b)

(c)

(d)

Fig. 6 The classification of the four attack types: **a** denial of service, **b** probe attacks, **c** privilege escalation attacks and **d** access attacks

4.4 Data Analysis

Step 1: A cross tab has been used to get the attacks on a few protocols icmp, tcp, udp.

Step 2: A helper function called bake_pies() was used to draw out multiple pie charts showing the attack in the network by using protocol as the parameter. They have been compared and analysed in Sect. 6.

Step 3: Helper functions called plt.show() and bake_pies() were used to draw out multiple pie charts showing the number of services in the attack sets "normal" and "attack" by taking flag as the parameter.They have been compared and analysed in Sect. 6.

Step 4: Helper functions called plt.show() and bake_pies() were used to draw out multiple pie charts showing the number of services in the attack sets "normal" and "attack" by taking service as the parameter. They have been compared and analysed in Sect. 6.

4.5 Feature Engineering and Building the Training and the Testing Sets

The initial set of encoded features were obtained and encoded. All the features were not there in the test set, so we accounted for the differences too. Few features like protocol_type, flag and service were showing a lot of variations. So, they needed to be brought to a base level for better identification. The training and the testing sets were built with 40% of the data for training and 60% for testing for both the two-class binary classification of attacks and the five class multi classification of attacks.

4.6 Building Out the Predictive Models

Models for binary classification and multi classification were made. The list of models (Random Forest Classifier, K-Neighbours Classifier, Logistic Regression) that we to be implemented for testing the data were defined in a list. The model was run and the performance of each model was calculated and captured in a list.

5 Comparative Analysis and Model Evaluation

5.1 Comparative Analysis of Data

During Data analysis attacks on the different protocols icmp, udp, tcp were analysed and pie charts were made. A observed that most of the attacks were targeted to a specific protocol tcp. Also, icmp protocol was the least frequently found protocol in the normal traffic.

Similarly, flag and service were also taken as a parameter to analyze the attacks in the network. It was observed that a huge number of the normal traffic is http and the attack traffic is everywhere.

Figures 7, 8 and 9 illustrate the pie charts generated for analysis of the attacks on the basis of protocol, flag and service as a parameter respectively.

5.1.1 Analysis of Attacks on the Basis of Protocol as a Parameter

Before the step of Feature Engineering, it is imperative to know the features which show a lot of variance. When, protocol is used as a parameter to categorize our data, it shows that the attacks are mostly attacking the TCP (Transmission Control Protocol) protocol. From the pie charts, it can be concluded that attacks like ipsweep, nmap, buffer_overflow are cross protocol attacks since ipsweep attacks both ICMP (Internet Control Message Protocol) and TCP protocol and NMAP (Network Mapper), buffer_overflow occurs in all the three routing protocols. This further predicts that TCP protocol is prone to most of the intrusion attacks (nearly all) followed by ICMP and then UDP(User Datagram Protocol). Therefore, most of the attacks target the TCP protocol with the buffer_overflow attack and the back attack

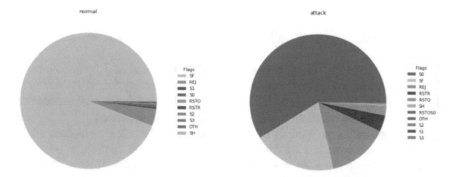

Fig. 7 Pie charts generated for analysis of the attacks on the basis of protocol as a parameter

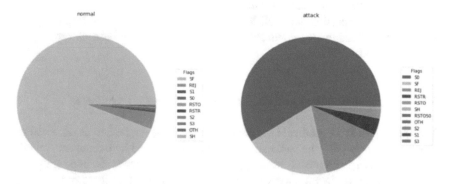

Fig. 8 Pie charts generated for analysis of the attacks on the basis of flag as a parameter

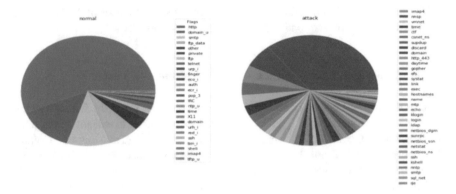

Fig. 9 Pie charts generated for analysis of the attacks on the basis of service as a parameter

being the most frequently occured and vulnerable ones. Buffer_overflow attacks target the UDP and ICMP protocols too but the frequency of occurrence is higher in TCP protocol.

5.1.2 Analysis of Attacks on the Basis of Flag as a Parameter

Here, flag refers to the feature 'attack_flag' which were marked 0 for normal and 1 for attacks. There are several connection statuses flags namely SF, REJ, RSTO, S1, S2, which vary in their occurrences in both the attack and the normal traffic. Few conclusions can be drawn from analyzing the pie charts generated:

In normal traffic, probability of occurrence of the SF's flag is highest which means that there is normal establishment and termination of connection occurring there. REJ flag's occurrence is also marked indicating that connection is attempted to be made but it has been rejected. In attack traffic, it is observed that S0 flag is mostly prevalent followed by SF and REJ flags, which means that most of the times, there has been an attempt to make the connection but there has been no reply from the Smart Sensor Network. Other marked occurrences are of RSTR flag that indicates that connection is established but has been aborted by the responder, and RSTO flag indicating that connection is established but has been aborted by the originator itself.

5.1.3 Analysis of Attacks on the Basis of Service as a Parameter

The application layer protocols act as an interface between the end devices and the smart sensor network. The application layer is implemented by the browser and the browser in turn implements these application layer protocols viz http, smtp, ftp, DNS, telnet. From the pie charts generated, it is observed that these protocols are prone to most of the Intrusion attacks in the Smart Sensor nodes. There are a huge number of attacks which target the http which is mostly vulnerable to the attacks occurring in the Smart Sensor Networks. Other than that, there is an array of services (application layer protocols) in the attack traffic.

5.2 Model Evaluation

After the training and the testing of the model, evaluation was done on the basis of these metrics.

(a) *Confusion Matrix*: A confusion matrix is a table that is used to describe the performance [16] of the model on a set of test data for which the true values are known. It is a tabular representation of the Actual vs the Predicted values. It is known as an error matrix where we have parameters like False Positive (FP), True Positive (TP), False Negative (FN), and True Negative (TN) for various classes.

The confusion matrix of the binary classification using Random Forest, Logistic Regression and K Nearest Neighbour as the Classifier are illustrated in Fig. 10a–c respectively. It was observed that Random Forest Classifier gave the best accuracy 99.3% among the three Supervised Learning Algorithms.

Figure 11 shows the confusion matrix for the multi classification using Normal, DoS, Probe, Privilege and Access as the classes using the Random Forest model.

(b) *Accuracy*: It is one of the evaluation metrics for evaluation of the classification models. It is the sum of True Positives and True Negative compared to the sum of True Positive, True Negative, False Positive and False Negative. Equation 1 depicts the formula for calculating accuracy:

$$Accuracy = \frac{\text{True Positive} + \text{True Negative}}{\text{True Positive} + \text{True Negative} + \text{Flase Positive} + \text{Flase Negative}} \quad (1)$$

(c) *Precision*: It yields the most predictive value. It is the number of True Positives compared to the sum of True Positive and False Positive. Equation 2 depicts the formula for calculating precision.

$$Precision = \frac{\text{True Positive}}{\text{True Positive} + \text{Flase Positive}} \quad (2)$$

(d) *Recall*: It is yielding the actual positive rate. It is the number of True Positives compared to the sum of True Positive and False Negative. Equation 3 depicts the formula for is calculating recall.

$$Recall = \frac{\text{True Positive}}{\text{True Positive} + \text{Flase Negative}} \quad (3)$$

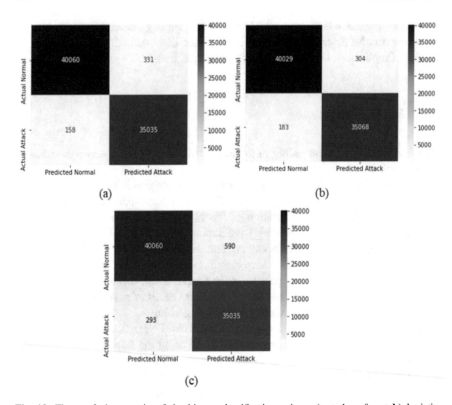

(a)　　　　　　　　　　　　(b)

(c)

Fig. 10 The confusion matrix of the binary classification using: **a)** random forest **b)** logistic regression and **c)** K-nearest neighbour algorithms

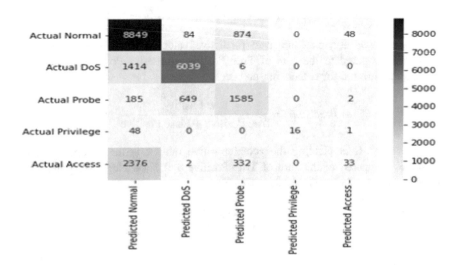

Fig. 11 The confusion matrix of the multi classification using random forest algorithm

Table 8 Accuracy, precision, recall and F1 score for the binary classification of attacks by using random forest, logistic regression and K-nearest neighbour algorithms

Evaluation metrics	LR	RF	K-NN
Accuracy	0.9833	0.993	0.982
Precision	0.98	0.99	0.98
Recall	0.98	0.99	0.98
F1 Score	0.98	0.99	0.98

Table 9 Accuracy, precision, recall and F1 score for the multi classification of attacks (i.e. normal, DoS, probe, privilege and access used as the classes) by using random forest, logistic regression and K-nearest neighbour algorithms

Evaluation metrics	LR	RF	K-NN
Accuracy	0.971	0.992	0.973
Precision	0.97	0.99	0.97
Recall	0.97	0.99	0.97
F1 Score	0.97	0.99	0.97

(e) *F1 score*: It is also a parameter used to measure the model's performance. It portrays a balance between the Precision and Recall. Equation 4 depicts the formula for is calculating the F1 Score.

$$F1\ Score = \frac{2 * True\ Positive}{2 * True\ Positive + False\ Positvie + Flase\ Negative} \quad (4)$$

The Accuracy, Precision, Recall and F1 Score for the binary and multi classification of attacks by using Random Forest, Logistic Regression and K-Nearest Neighbor Algorithms are depicted in Tables 8 and 9 respectively.

6 Performance Analysis

From the confusion matrices, Random Forest (RF) Algorithm was found to be the best technique for this work. RF classified every class the most accurately. The confusion matrix of KNN is similar to RF. 304 samples were flagged as attacks which were actually normal. For the multiclass classification by Random Forest Classifier, 8849 samples were predicted as normal which were also actually normal, 6039 samples were predicted having DoS attack which were also actually having DoS, 1585 samples were predicted having Probe attacks which actually had Probe attacks, 16 actual privilege samples were flagged as privilege, 335 samples were predicted to have Access attacks also fell under 'Actual Access'.

The proposed model achieved a very high level of accuracy showing that the model is a successful method for the detection of external threats on the Smart

Sensor Networks. It can be deployed in the smart sensor network ensuring their security against harmful threats. This chapter provides a much more detailed description of the dataset and a clear explanation of the data pre-processing which is vital before creating any Machine Learning Model. This chapter also focusses on detecting the different classes of attacks which is even more challenging than binary classification. Both of them are depicted here. Finally, the Model Evaluation depicted according to the Confusion Matrix, Accuracy, Precision, Recall and F1 Score shows that our model has performed better than other existing threat detection models.

7   Future Research Directions

The biggest challenge of these smart sensor networks is the external threat detection which is a rising concern in the field of smart sensor network in every domain. In the IoT Network, micro services behave differently at different times which cause deviations from normal behavior in IoT services creating an anomaly. Further research is needed to interpret this problem. A new algorithm can be developed which focuses on real time data so that there may arise different problems and they can be solved. The major benefit from real time Big Data analytics for security of the smart sensor networks is condition based maintenance. It includes managing the performance of the models to reduce the likelihood of failures and to check if the system needs to be serviced. By the help of a new algorithm which will focus on real time data, condition-based maintenance can be done smoothly.

Deep learning can be used for the attack detection in the Smart sensor networks to make more accurate predictions of cyber-attacks occurring in the Smart sensor nodes. With the help of this, the attacks can be mitigated beforehand before much damage is caused to the system. There are different areas of Deep learning which run only on real time data sets like reinforced learning or Neural networks which can be implemented to make the system capture attacks automatically and mitigate them before the smart sensor networks is crashed.

8   Conclusion

The main objective of this paper was detection of threats occurring in IoT applications and classification of attack types in detail for an in-depth understanding of the attacks. The biggest challenge of these smart sensor networks is the external threat detection which is a rising concern in the field of smart sensor network in every domain. DoS Attacks, Probe attacks, Privilege Escalation Attacks and Random-Access attacks have been studied and explained in detail. Three popular Supervised Machine Learning Classification Algorithms have been used to classify and detect the external threats occurring in the smart sensor networks.

While evaluating the system, we have taken help of confusion matrices, accuracy, precision, recall and F1 score parameters to compare the accuracy of the predictions. The experiment demonstrated that one should consider using Random Forest Classifier on these kinds of datasets for the detection of cyber-attacks on Smart Sensor Networks because the proposed model predicted the attacks more accurately with RF classifier as compared to the other classifiers. Thus, this model can be deployed to secure the smart sensor networks against harmful threats thereby keeping a check on the sensor network and preventing any system failures or a crash of the entire system.

References

1. Hassija, V., Chamola, V., Saxena, V., Jain, D., Goyal, P., Sikdar B.: A survey on IoT security: application areas, security threats, and solution architectures. IEEE Access (2019)
2. Jindal, M., Gupta, J., Bhushan, B.: Machine learning methods for IoT and their Future Applications. International 23 (2019)
3. Proceedings of International Conference on Artificial Intelligence and Applications. Springer Science and Business Media LLC (2021)
4. Albanesi, M.G., Ferretti, M., Guerrini, F.: A taxonomy for image authentication techniques and its application to the current state of the art. In: Proceedings 11th International Conference on Image Analysis and Processing, pp. 535–540. IEEE (2001)
5. Novo, O.: Blockchain meets IoT: an architecture for scalable access management in IoT. IEEE Internet Things J. 5(2), 1184–1195 (2018)
6. Hassija, V., Chamola, V., Saxena, V., Jain, D., Goyal, P., Sikdar, B.: A survey on IoT security: application areas, security threats, and solution architectures. IEEE Access 7, 82721–82743 (2019)
7. Canedo, J., Skjellum, A.: Using machine learning to secure IoT systems. In: 2016 14th Annual Conference on Privacy, Security and Trust (PST), pp. 219–222. IEEE (2016)
8. Aghajan, H., Augusto, J.C., Wu, C., McCullagh, P., Walkden, J.A.: Distributed vision-based accident management for assisted living. In: International Conference on Smart Homes and Health Telematics, pp. 196–205. Springer, Berlin, Heidelberg (2007)
9. Arth, C., Bischof, H., Leistner, C.: Tricam-an embedded platform for remote traffic surveillance. In: 2006 Conference on Computer Vision and Pattern Recognition Workshop (CVPRW'06), pp. 125–125. IEEE (2006)
10. Alzaid, H., Foo, E., Gonzalez Nieto, J.M.: Secure data aggregation in wireless sensor networks: a survey. In: Proceedings of the Sixth Australasian Information Security Conference (AISC 2008), vol. 81, pp. 93–105. Australian Computer Society (2008)
11. Shahid, L., Zou, Z,. Idrees, Z., Ahmad, J.: A novel attack detection scheme for the industrial internet of things using a lightweight random neural network. IEEE Access (2011)
12. Akyildiz, I.F., Melodia, T., Chowdhury, K.R.: A survey on wireless multimedia sensor networks. Comput. Netw. 51(4), 921–960 (2007)
13. Aaraj, N., Raghunathan, A., Jha, N.K.: Analysis and design of a hardware/software trusted platform module for embedded systems. ACM Trans. Embed. Comput. Syst. (TECS) 8(1), 1–31 (2009)
14. Holding, A.R.M.: ARM Security Technology, Building a Secure System using TrustZone Technology (2009)
15. Machine Learning, Image Processing, Network Security and Data Sciences. Springer Science and Business Media LLC (2020)

16. Koroglu, Y., Sen, A., Kutluay, D., Bayraktar, A., Tosun, Y., Cinar, M., Kaya, H.: Defect prediction on a legacy industrial software. In: Proceedings of the 4th International Workshop on Conducting Empirical Studies in Industry—CESI '16 (2016)
17. ElMamy, S.B., Mrabet, H., Gharbi, H., Jemai, A., Trentesaux, D.: A survey on the usage of blockchain technology for cyber-threats in the context of industry 4.0. Sustainability (2020)
18. Da Xu, L., He, W., Li, S.: Internet of things in industries: a survey. IEEE Trans. Industr. Inf. 10(4), 2233–2243 (2014)
19. Zou, Y., Zhu, J., Wang, X., Hanzo, L.: A survey on wireless security: technical challenges, recent advances, and future trends. Proc. IEEE 104(9), 1727–1765 (2016)
20. Mpitziopoulos, A., Gavalas, D., Konstantopoulos, C., Pantziou, G.: A survey on jamming attacks and countermeasures in WSNs. IEEE Commun. Surv. Tutorials 11(4), 42–56 (2009)
21. Advances in Soft Computing and Machine Learning in Image Processing. Springer Science and Business Media LLC (2018)
22. Zhou, L., Wu, D., Zheng, B., Guizani, M.: Joint physical-application layer security for wireless multimedia delivery. IEEE Commun. Mag. 52(3), 66–72 (2014)
23. Zhang, L., Restuccia, F., Melodia, T., Pudlewski, S.M.: Learning to detect and mitigate cross-layer attacks in wireless networks: framework and applications. In: 2017 IEEE Conference on Communications and Network Security (CNS), pp. 1–9. IEEE (2017)
24. Ashton, K.: That 'internet of things' thing. RFID J. 22(7), 97–114 (2009)
25. João, R., Amorim, M., Cohen, Y., Rodrigues, M.: Chapter 23 Artificial Intelligence in Service Delivery Systems: A Systematic Literature Review. Springer Science and Business Media LLC (2020)

Towards Smart Farming Through Machine Learning-Based Automatic Irrigation Planning

Asmae El Mezouari, Abdelaziz El Fazziki, and Mohammed Sadgal

Abstract The growing scarcity and strong demand for water resources require an urgent policy of measures to ensure the rational use of these resources. Farmers need irrigation planning and rationalization tools to be able to take advantage of scientific know-how, especially artificial intelligence tools, to improve the management of water use in their farming irrigation practices. To improve water management in irrigated areas, models for estimating future water needs are needed. The objective of this work is to estimate the water needs of crops for efficient management of irrigation networks and planning of the use of hydraulic resources. In this regard, data-driven machine learning algorithms can be employed for water resources monitoring and governance. These methods, derived from artificial intelligence, have obtained promising results in the planning, management, and control of water resources. To do this, we prepare a dataset with information about the appropriate attributes for calculating water requirements. The proposed approach begins with a cleaning of the data set to effectively predict water needs. The process of extracting relevant data is based on a combined tool for data mining and knowledge discovery on irrigation and water needs. We then validate the effectiveness of the various data mining algorithms used and of certain traditional methods of estimating evapotranspiration (ETc) to predict water requirements, in particular the Water balance (WB), the Penman-Monteith method (FAO PM) adopted by the Food and Agriculture Organization of the United Nations, and the Bowen-Energy Balance Report (BREB). Some of the algorithms used include XGBoost, Random Forest, and Deep Artificial Neural Networks. Currently, innovations can be consolidated to minimize costs and maximize the use of resources.

A. E. Mezouari (✉) · A. E. Fazziki · M. Sadgal
LISI, Faculty of Technical Sciences, FST-UCAM, Marrakech, Morocco
e-mail: asmae.elmezouari@ced.uca.ma

A. E. Fazziki
e-mail: elfazziki@uca.ac.ma

M. Sadgal
e-mail: sadgal@uca.ma

© The Author(s), under exclusive license to Springer Nature Switzerland AG 2022 179
U. Singh et al. (eds.), *Smart Sensor Networks*, Studies in Big Data 92,
https://doi.org/10.1007/978-3-030-77214-7_8

Keywords Smart farming · Machine learning · Water balance · FAO
Penman-Monteith · Bowen-Ratio-Energy balance

1 Introduction

Regarding the fast increase in water use for farming and population activities, appropriate and effective water resources management practices and policies are urgently needed and should aim to ensure the satisfaction of the water for the populations. Crop water requirements vary in space and time depending on weather conditions, the crop's type, and the crop's growth stage. Since the advent of New Technologies and Innovative Techniques, the quality of agricultural products has continued to deteriorate. Often, the majority of people are oblivious to the right time and the right place for culture. Due to these cultivation techniques, seasonality, climatic conditions are also changed concerning basic assets like soil, water, and air which lead to food insecurity. By analyzing all the parameters that come into play for better crop yield, there is no suitable solution and technique to overcome the situation with which the users of water resources are confronted. In India there are several, there are many ways to increase and improve the yield and quality of crops. The analysis and extraction of data indicators have become essential to predict water needs and crop production. Typically, data mining involves data analysis from different sides and extract relevant information from it. Analysis and data mining tools allow users to perceive data in different dimensions, classify it, and define its correlation. Technically, data analysis and forging are the process of finding correlations or modeling parameters in very large data sets. The models, associations, or relationships between all these data allow knowledge extraction in the field through automatic learning from historical data and define future trends. For example, analyzing and synthesizing information on agricultural production can help farmers identify the causes of a poor harvest, poor yield, and predict the future. Since forecasting crop yields is an interesting agricultural problem, every farmer is always eager to know what yield he will get, whatever the circumstances. In the past, the forecast of the yield or the water needs was projected by analyzing the previous experiences on a particular crop. Crop yield mainly depends on weather conditions, water availability, and planning for irrigation and harvesting operations. Accurate information on the history of water needs is important for making decisions related to agricultural risks.

Therefore, this work proposes an approach to predict the water needs of a crop. This approach is based on artificial intelligence tools, more particularly supervised machine learning algorithms, and a combination of knowledge on irrigation and data mining tools. The rest of this paper is fluently structured as follows. Firstly, Sect. 2 exposes a concise literature review. Then, Sect. 3 expresses smart farming and Irrigation Scheduling. Then, Sect. 4 portrays the different functionalities of the proposed framework. Then, Sect. 5 illustrates the proposed approach with a case study of automatic corn irrigation planning. Finally, this work ends with a brief conclusion and some perspectives.

2 Related Works

The need for irrigation water for agriculture is a matter of major concern, especially given the growing demand and availability of resources which is constantly decreasing, and the need to compete with other consumption sectors [1]. In the last decade, multifarious researches had been made in intelligent decision systems development in water resources management, in particular smart irrigation systems. Which have seen an extensive evolution benefiting from the Internet of Things revolution, Big Data, and Machine Learning. Some of them are specific in predicting soil and climatic features like evapotranspiration, by analyzing time series collected from some past years to compare three types of evapotranspiration estimation methods. Data entrees Resampling techniques and diverse machine learning algorithms like (Support Vector Regression, M5P Regression Tree, Bagging, and Random Forests) have been applied using data from an irrigated site in Central Florida, that has humid subtropical weather. Additional research works in [2], have done analysis comparing two machine learning streamflow modeling to predict streamflow through the TUW hydrological model, Extreme Gradient Boosting (XGBoost), multiple linear regression (MLR), Deep Learning Neural Network (DLNN), and Random Forest (RF) techniques. In particular, one employed climatic data (precipitation, temperature, and potential evapotranspiration), while the other also introduced the past progress in the data entrees. The evaluation of these models was constructed using statistical analyzes of the standard mean squared error (RMSE), coefficient of determination (R2), Kling-Gupta efficiency (KGE), Nash-Sutcliffe efficiency (NSE), and percent bias between simulated and observed measures (PBIAS). Improving the accuracy of flow simulation techniques has included two options, one concerning the impact of the selected approach on the precision, the other concerning the effect of applying engineering characteristics for flow modeling, based on the addition of Variables selected from historical and climatic data. To verify the precision and the efficiency of the predictive approaches, a comparative study was done on a variety of datasets between XGBOOST and traditional algorithms such as K-Neighbors Classifier (KNN), Linear Discriminant Analysis (LDA), Support Vector Machines (SVM), Decision Trees (DT), AdaBoost (AB), Naïve Bayes (NB), Logistic Regression (LR) and Random Forest (RF). Therefore, the results maintain that XGBoost is truly the best approach because its given results are more accurate (100% accuracies) than its equivalent models on all the datasets referring to [3]. An in-depth recent literature study is carried out in [4] to expose and analyze the latest discoveries in smart farming systems, with an emphasis on measuring the impact of artificial intelligence techniques on improving productivity in agriculture and optimizing water use in irrigation. Also, this survey sheds the light on the equipment role of the object's

interest in the improvement of the irrigation procedures by facilitating the acquisition of the information of the cultures and the ground. Also, the various machine
learning algorithms such as the partial least square regression (PLSR), the adaptive
neuro-fuzzy inference system (ANFIS), the artificial neural networks (ANN), and
many other tools are used to compare the pertinence of several input parameters to
evaluate their impact on the accuracy.

Evapotranspiration (ET) estimation is at the queen's interest in the hydrologic
cycle studies but it is still on the line to uncertainties. Hence, Estimates and predictions of the ET constitute an essential step of irrigation management over the
world. In this context, a multitude of studies has been made to improve the ET
Estimation. Especially, in [5] the study used Bowen ratio and eddy covariance
methods for calibrating, extracting, and calculating relevant parameters necessary
for guessing ET of tea canopy for the whole growing season, to provide further
predictions about the adoption of this method for scheduling other crops irrigation
as well. Also, another comparative study performed a comparison between three ET
estimating methods, in particular the eddy covariance (EC), the Bowen ratio-energy
balance (BREB), and the soil water balance (SWB) to measure their efficiency
during the cropping season of corn in [6]. Moreover, The FAO Penman-Monteith
(FAO PM) became a specification for ETo estimation over the last decades. This
method takes into account many climatic variables linked to the evapotranspiration
activity such as the net radiation (Rn), the air temperature (T), the vapor pressure
deficit (Δe), and the wind speed (U); and its results are very satisfactory referring to
[7]. Besides, over the last few decades, multifarious observational methods have
been developed by many scientists and specialists around the world to estimate
evapotranspiration based on climatic variables. However, Relationships used in
these approaches have been a subject of severe local calibrations. Even though, they
remain more or less limited in terms of validity according to the review of various
ET methods; which indicates how to apply the FAO PM as a consistent and all-over
valid uniform for crop-water-requirement estimation, in the case when the available
climatic data are missing or limited. Hence the need to test the accuracy of these
methods under different conditions in our comparative study, and eventually to use
the data preprocessing techniques for an accurate Evapotranspiration estimation for
better irrigation planning. Also, after much experimentation and research, it appears
that real-time monitoring and control was not effective enough to achieve better
management of water resources concerning irrigation in agriculture. Yet a predictive approach in the short or long term will be the magic solution for predicting the
soil state and the future availability of water to act more accurately benefitting from
advanced based on the algorithms of machine learning derived from Artificial
Intelligence theories.

A smart crop irrigation system is often a long-term solution. Therefore, we need
to have solutions that can be scalable according to changes in the crop environment
such as in our case. In addition, IoT and Machine Learning algorithms like Deep
Learning, Random Forest, and XGBoost allow classification and predictions related
to different soil and crop type, and water requirements as well. Thus, most automatic irrigation systems proposed are based on a single algorithm and a single

method of calculating water requirements. A comparative analysis using various machine learning algorithms and various techniques for calculating water requirements provides several alternatives that adapt to the context of the problem and the quality of the data used. According to the modest experience we have conducted, deep learning is appropriate for large masses of data while other algorithms are more suitable for small datasets. The benefit of this approach is that it can be implemented in parallel in real-time in a huge distributed environment such as HADOOP/MapReduce which is open source and very performant according to [8] and to [9].

3 Smart Farming and Irrigation Scheduling

3.1 Smart Farming

Smart farming improves yields by using minimal resources such as water, fertilizers, and seeds. Farmers can easily deploy sensors to remotely monitor their crops, conserve resources, and reduce the impact of climate changes on crops [10]. Several parameter detection technologies are used in this agriculture for providing data and helping farmers to monitor and optimize their crops [11], as well as to adapt to environmental change factors, including location, electrochemical, mechanical, airflow sensors, agricultural weather stations, humidity, and PH. Figure 1 illustrates the factors that influence the irrigation system.

One of the main sensors in smart farming is that of soil moisture. It's used for measuring the present soil volumetric water content (humidity). The threshold value is fixed, and the soil moisture level is measured and verified with the upper and lower thresholds at the necessary levels. Irrigation is the vital need of agricultural activities, there are three classic irrigation methods of which we can cite canal irrigation, sprinkler irrigation, and drip irrigation responding to the needs of plants,

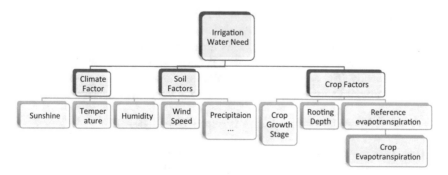

Fig. 1 The most important irrigation planning features

these three methods are used. Regarding the intelligent irrigation system, the researchers in [12] have shown that water consumption is minimized when an automated irrigation system relies on soil moisture as an implementation parameter. Among these irrigations, that of drip is the one where farmers can save the most water because it will provide water in the form of droplets directly on the plant root and the soil surface.

IoT-based intelligent irrigation management systems can help achieve optimal use of water resources in the precision agriculture landscape. This paper presents an intelligent system mainly based on open-source technology; to help farmers in scheduling irrigations, by predicting irrigation needs of the field using the detection of soil parameters, such as the moisture and the temperature in the soil, and the environmental states, as well as the internet weather forecast data. Detection nodes, involved in soil and environmental detection, take into account soil moisture, soil temperature, air temperature, ultraviolet radiation, and relative humidity of the culture soil. The intelligence behind the proposed system comes from the fact that the intelligent algorithms employed take into account the detected data as well as the weather-predicted parameters such as precipitations, air temperature, humidity, and ultraviolet radiation. Due to a lack of means and data collection devices, the proposed system has been developed and deployed at a validation scale. Where the data measured by the sensors is historical and the water requirements are predicted using machine learning algorithms based not only on the analysis of the sensor data but also on the forecast weather data. The system is intended for use in the closed-loop control of the water supply to achieve a fully autonomous irrigation system. This work describes the proposed system and discusses in detail the results obtained by the different algorithms and the different methods of calculating evapotranspiration. The system provides good predictive models and the forecast results are very encouraging.

3.2   The Water Balance Approach

To the greatest extent, the net irrigation need is designated as the volume of water mandatory to fill the moisture of the root zone up to the saturation of the field related to some researches in [13]. This quantity is called the soil water deficit (D) and conforms to the deviation between the field capacity and the current level of the soil humidity. The expert can keep a record of the net water irrigation amount to apply D, through Daily estimation using the following water balance equation:

$$Dc = Dp + ETc - P - Irr - U + SRO + DP \tag{1}$$

where Dc is the current day's water deficit, P is the current day's gross precipitation, ETc is the current day's crop evapotranspiration's rate, Dp is the previous day's soil's water deficit, Irr is the irrigation volume percolated into the soil for the same day, ORS is the surface runoff, DP is the deep percolation, and U is an influx of shallower groundwater infiltrated into the root depth.

The last three variables of the previous equation (U, SRO, DP) are the most complex to estimate compared to the other features in the field. In a multitude of cases, the water table is very deep and far from the root zone. Hence U and SRO are systematically equal to zero. Also, Dc could be simply negative and replaced by zero each time the infiltrated water to the root zone from precipitation and irrigation exceeds the deep percolation and the evapotranspiration (P + Irr > Dp + ETc) which indicates that the soil is saturated. Based on these facts, Eq. 1 could be optimized to:

$$Dc = Dp + ETc - P - Irr \qquad (2)$$

3.3 The Bowen Ratio-Energy Balance

The BREB is considered an efficient and successful method for estimating ET, which builds essentially upon the energy balance between the gaseous properties of water- vapor and the surface energy fluxes related to [14]. Bowen ratio presents the ratio of the tangible heat flux (H) to the potential heat flux (LE) which accomplishes within the concept of tangible and potential heat diffusion.

$$\lambda E = \frac{R_n - G}{1 + \beta} \qquad (4)$$

Where λE is the Potential evapotranspiration (mmday-1), G is the Soil heat flux (MJday-1), Rn is the Net radiation (MJday-1), and β is the Bowen ratio which is even calculated as the formula below:

$$\beta = \lambda \frac{\Delta T}{\Delta e} \qquad (4)$$

Where λ is psychometric constant, ΔT is temperature change, and Δe is vapor-pressure change.

3.4 The FAO PM Method

Referring to [15], The FAO Penman-Monteith (FAO-PM) equation is adopted by experts as the standard method for ETo estimation based on weather data because of its performance revealed in several evaluation studies.

The energy balance and the mass transfer method were combined by Penman in 1948. Then an equation for computing evaporation from an open water surface

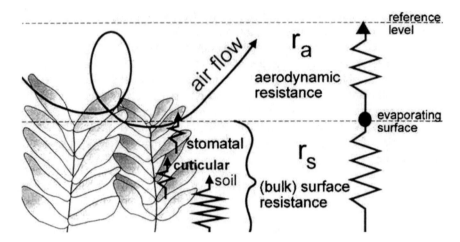

Fig. 2 The field resistances' scheme of the water vapor flow [16]

based on standard meteorological data in particular sunshine, temperature, humidity, and wind speed. Later, further developments have been derived from this combination by many researchers to introduce resistance factors to adopt it for cropped surfaces.

Figure 2 portrays the distinctions between aerodynamic resistance and surface resistance factors expressed by the resistance nomenclature.

The 'bulk' surface resistance parameter which works sequentially with the aerodynamic resistance is a combination of surface resistance parameters. The resistance of the vapor flow passing through the openings of the stomata, the soil surface, and the total leaf area are described by the surface resistance, rs. also, the resistance of the vegetation upwards and the friction of the air circulating on the vegetative surfaces are grouped in aerodynamic resistance, ra. despite the complexity of describing the exchange process in the flora, at least the correlations obtained between the evapotranspiration rates remain good, eventually for uniform herbaceous reference surface.

Penman-Monteith's combination equation is:

$$\lambda ET = \frac{\Delta(R_n - G) + PaCp\frac{Es-Ea}{Ra}}{\Delta + \lambda\left(1 + \frac{Rs}{Ra}\right)} \tag{5}$$

Where G is the soil heat flux, Rn is the net radiation, (Es − Ea) is the air's vapor pressure's deficit, λ is the psychometric constant, Cp is the air's definite heat, Ra is the mean of air's density at constant pressure, Δ is the saturation' slope vapor pressure-temperature relationship, and Ra and Rs respectively represent the aerodynamic and the (bulk) surface resistances.

The Penman-Monteith equation employs all features that englobe energy exchange and evapotranspiration ensue from stable ranges of plants. The majority of these variables can be covered or can be simply assessed using weather data. Knowing that both surface and aerodynamic resistances are specific for each crop, the equation can then be employed for specific evapotranspiration estimation of any crop.

The heat transfers and water vapor from the evaporating surface into the air above the canopy are expressed by the aerodynamic resistance and calculated using the formula:

$$Ra = \frac{\ln\left[\frac{Zm-d}{Zom}\right]\ln\left[\frac{Zh-d}{Zoh}\right]}{k^2 Uz} \tag{6}$$

Where Ra is the aerodynamic resistance [sm^{-1}], Zm is the wind's height [m], Zh is humidity 's height [m], d is the zero-plane displacement's height [m], Zom is the roughness' length controlling momentum transfer [m], Zoh is the hole length prevailing the transfer of heat and vapor [m], k is the von Karman's constant [−], and Uz is the wind speed at height z [m s^{-1}].

This restricted equation is designed to be used only for the neutral stability conditions, where climatic variables distributions (atmospheric pressure, wind velocity, and temperature) go nearly along with adiabatic conditions (without heat exchange). Thus, applying this formula for a brief duration (hourly or Instantaneous, or less) might need modifications for keeping the stability. However, whenever the prediction of ETo is made in a well-watered reference surface, we can deduce that the heat exchanged is small, and consequently, stability correction isn't any more necessary.

The "Bulk" surface resistance refers not only to the resistance to the flow of vapor within the transpiring crop but also to the surface evaporation. whenever the soil isn't completely covered with the vegetation, the evaporation effects from the soil surface should be included in the resistance factor as a result. The resistance relies also on the vegetation's water level in the case when the crop transpiration isn't potential.

$$Rs = \frac{rl}{LAI\ active} \tag{7}$$

Where rs (bulk) is the surface resistance [s m-1], rl is the bulk stomatal resistance of the well-illuminated leaf [s m^{-1}], and LAI active (sunlit) is the leaf area index [m^2].

The global stomatal resistance rl is denoted by the mean resistance of a single leaf. This measure depends on the crop treatments and variability. It often rises with urbanization and the rip of crops. However, information about fluctuations in rl over time for the various plants is usually incomplete.

3.5   *Water Needs Prediction Process*

To guarantee potential yields, the water requirements in the root zone of a crop must be met. These needs are generally designated by crop evapotranspiration and represented by ETc. Evapotranspiration combines the soil surface evaporation or the plants' wet surfaces, and the perspiration of leaves. Water requirements could be served by precipitation, underground water, or irrigation. Irrigation is therefore necessary when the crop water need (ETc) outstrips both the water stored and the precipitations. Knowing that the ETc depends on the crop stage and meteorological changes, the quantity and the period of irrigation are critical. The water balance approach allows easy estimation of water requirements for irrigation planning. This method as mentioned before in the equation is based on several factors particularly, the initial water content of the soil in the root zone, the ETc, the precipitation, and the soil's capacity.

The daily Crop evapotranspiration (ETc) is accessed as:

$$ETc = ETr \times Kc \times Ks \tag{8}$$

Where ETr is the reference crop evapotranspiration rate, Kc is the crop coefficient that depends on the development stage (0 to 1), and Ks is the water stress coefficient (0 to 1). For each stage in the growing season, the Kc for each crop is basically estimated as the ratio of its ETc over the reference crop ETr [13].

ETr (reference evapotranspiration) is estimated on the basis of climatological variables and expresses the effect of meteorological conditions on the net water requirements of crops, while Kc marks characteristics of the crop and its effect on the plant requirements. water (the type of culture, development, and phonological stage,

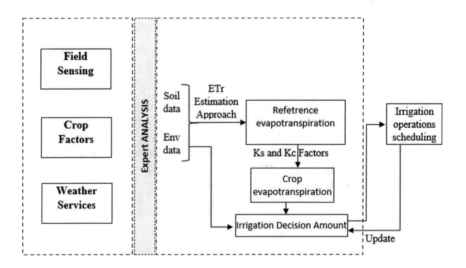

Fig. 3 The expert irrigation planning process

etc.). Thirdly, the quality of the irrigation water, the uniformity coefficient of the irrigation system, the size of the field, etc., make it possible to determine the actual irrigation amount. This value gives a rough idea of the volume of water needed to meet the crop's water needs. This value is balanced with the hydric state of the soil to obtain the volume of irrigation for contribution to the harvest stage (Fig. 3).

4 The Proposed Framework

4.1 An Overview

The main goal of this framework is to calculate the irrigation water needs that must be applied to crops. The purpose of this component is therefore to imitate an expert (agronomist) in the process of determining the water requirements for a crop. The main parts used to develop the framework are described: an integrated information system that provides data on the soil, irrigation, crops, and plots where the irrigation data and reports were calculated and tested by experts in the field.

The proposed system is a tool that predicts the amount of irrigation water needed in two depts. (105 cm and the root zone) for a crop field based on two crop evapotranspiration estimated by the experts using Water balance (WB), FAO Penman-Monteith (FAO PM), and Bowen Ratio-Energy Balance (BREB) approaches to compare their efficiency in planning irrigations. To cover the best options for machine learning processes, several machine learning techniques will be implemented in Python, in particular XGBoost, Random Forests, and Deep Neural Networks. Then a performance analysis is done based on the evaluation metrics below will be applied to select the best predictive model. The preprocessing of data to extract the relevant information using the principal component analysis method will be done too. Figure 4a illustrates the framework components and Fig. 4b details the processing architecture.

4.2 Decision-Making Process

The proposed decision-making process in Fig. 5 for crop irrigation management was developed based on root growth patterns, soil water balance, crop water production, crop type, and the irrigation management model. The irrigation plan is made by predicting the water content of the soil in the root zone and the daily water requirements of each crop using historical and forecast weather data. This process performs different irrigation estimates based on the optimization of water needs and historical experience and allows user personalized irrigation. The main function of this process includes a decision-making aid for managing irrigation in the different phases of the crop and taking into account soil dynamics in terms of water stored in the root zone. A case study of Corn crop irrigation management will be carried out to show the added value of the process and its practical benefits.

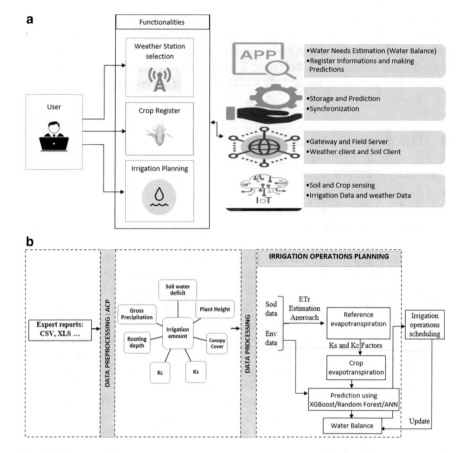

Fig. 4 The proposed system components

4.3 *Materials and Methods*

4.3.1 XGBoost

XGBoost is an ingenious learning model created by GBMs to give accurate solutions based on ensemble learning trees. The main function of XGBoost is to build weak learners that classify the data and make predictions based on incorporating the predictive power of a multitude of learners. Along with the iterations, the accuracy improves and the final output results are the aggregated predictions from diverse weak predictive models that reduce the previous models' errors according to [3]. XGBoost can be run on major distributed environments like Hadoop, SGE, and MPI and it has been implemented in various programming languages.

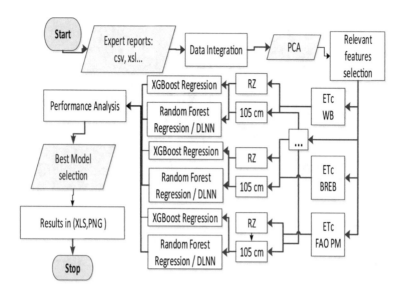

Fig. 5 Irrigation decision-making process

4.3.2 Random Forest

Random Forest is one of the Supervised Learning algorithms which uses a set of learning methods to perform classification and regression tasks. It combines several tree predictors. Each tree provides its prediction. At the processing end, the commonly chosen prediction is selected as a result such as the mode of the classes in classification and the mean prediction in regression referring to [17].

4.3.3 Deep Learning Neural Networks

Is the most exciting and powerful subdomain of Machine Learning, which is a collection of computing models that mimics the human brain in data processing to help in making decisions. They are recognized as Artificial Neural Networks (ANN) and inspired by the processing way of the biological nervous system [18]. The following diagram represents the general model of an ANN which is similar to the natural biological neuron organism. On denote Perceptron single-layer neural network referring to [19]. It gives a single output as shown in Fig. 6.

Both XGBoost, Random Forest and Artificial Neural Networks models have proved recently its accuracy and efficiency in many studies.

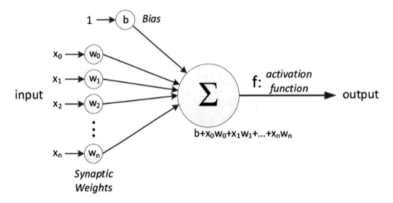

Fig. 6 The ANN perceptron structure

4.3.4 Evaluation Metrics

Given in the following equation, The Root Mean Square Error, the mean square error, the mean absolute error (MAE), and the r-squared accuracy will be adopted to measure the precision of the predictive models:

$$RMSE = \sqrt{\sum_{i=1}^{n} \frac{(yi - \bar{y}_i)^2}{n}} \qquad (9)$$

Where n is the number of data, yi is the actual output of instance i, and $\bar{y}_i$ is the appropriate resulting output. It is a very standard and global error measure where the more the value is lower, the more the forecast precision is higher. Note that the RMSE is measured on the analogous range as the output variable. In the fact, easy comparison between the RMSE of the predictive methods is sufficient to assess their performance when the output parameter is similar for all the predictive models.

Mean absolute error (MAE) is a measure of errors between two observations describing the same thing(measure). MAE is calculated as:

$$MAE = \sum_{i=1}^{n} \frac{|yi - xi|}{n} \qquad (11)$$

R-squared measures how much the data is near the adapted regression course. It may represent the determination coefficient for single regression or the multiple determination coefficient for multiple regressions:

$$R - squared\ accuracy = Explained\ variation/Total\ variation \qquad (11)$$

The R-squared score is expressed as a percentage between 0 and 100:

- 0 means that the variability interpretation in response data around its mean provided by the predictive model is null.
- 100 indicates that the predictive model provides very well all variability's interpretations in predicted data around its mean.

4.3.5 Description of Output and Input Variables

Regarding all the available training features, the choice of training input was done related to information used by agronomists during the irrigation estimates process, such as meteorological data necessary for estimating crop evapotranspiration, total water requirements, soil water status, the quantity of water previously applied, and the critical level of the harvest period: the main critical periods are flowering and setting (stage I), and the second period of plant's quick growth (stage II).

However, more factors could affect the forecasting of irrigation (such as prediction time, possible interruption of irrigation in the area, etc.), and these factors are not taken into account by the experts. This could be a limitation of this approach.

Selecting an appropriate set of features is decisive for the good performance of prediction models. In this sense, a technique for selecting the most relevant parameters will be used, denoted the Principal Component Analysis (PCA).

4.3.6 Principal Component Analysis

PCA is an analytical method employed in multivariate statistics, which consists of converting variables correlated with one another into novel variables unrelated to one another. These novel variables are called "principal components. It provides some technics to optimize the set of variables, to reduce the information redundancy, and to improve the relevancy. These variables will be labeled (Main components), the axes they determine (main axes), and the correlated linear forms (main factors) [20].

Related to [21] and [22] The objectives pursued by PCA are:

- The "optimal" graphic representation of individuals (lines), minimizing the deformations of the point cloud, in a subspace E of dimension q (q < p),
- The graphic representation of the variables in a subspace F by explaining at best the initial links between these variables,
- Reduction of the dimension (compression), or approximation of X by an array of rank q (q < p).

5 Case Study

5.1 Water Requirements for Grain Corn

Corn is a C4 plant (like sugar cane or sorghum), which gives it the particularity of having greater photosynthetic efficiency, that is to say, that it uses carbon dioxide better and produces more oxygen than C3 plants (most other crops). C4 plants need almost two times less water to absorb 1 g of carbon than C3 plants. The water requirements of corn depend on all the climatic factors of the day expressed in Potential Evapotranspiration, and they differ according to the stage of the plant.

The water requirements of maize depend on all the climatic factors of the day expressed in Potential Evapotranspiration (FTE), and they differ according to the stage of the plant presented in Fig. 7 and Table 1. From the 8-leaf stage (V2) to the mid-rise stage (V12), the need for water increases very quickly. Maize is at its maximum from the mid-stage to the limit stage of grain abortion. Then these needs are slightly lower while remaining high up to the 50–45% grain moisture stage. Beyond that, the needs decrease quite quickly. The pre-and post-flowering period is a stage particularly sensitive to lack of water.

In the present case study, we focus on estimating the daily water requirement of Corn during the last growing season between 05/14/2013 and 11/05/2013 containing almost 175 days using the water balance approach [24]. First, we analyze the main components of 28 parameters in the dataset presented in Table 2 to select the relevant set of parameters (12 features). Next, we predicted the parameters

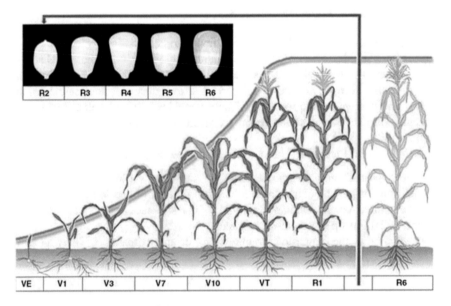

Fig. 7 The corn growth stages [23]

Table 1 The corn growth stages signification

Variable	Vegetative stages	Variable	Reproductive stages
–	Plant	R1	Silk
VE	Emergence	R2	Blister
V1	First leaf	R3	Milk
V2	Second leaf	R4	Dough
V(n)	nth leaf	R5	Dent
VT	Tassel	R6	Maturity (Harvest)

influencing irrigation in Fig. 8 by applying supervised machine learning to the dataset in Python in the Anaconda environment, over 4074 days collected and estimated daily by experts at the USDA-Agricultural Research Service. Finally, we estimate the amount of water needed for corn irrigation using the simplified Eq. (2) of the water balance. The dataset was gathered from a study on the relationship between water productivity and seasonal timing corn irrigation calendar at the Limited Irrigation Research Farm (LIRF) in northeastern Colorado. This dataset is of size (4249) records and collected for two years 2012–2013, which were full years with intact treatments. The database englobes the canopy's growth and development (canopy height, canopy cover, and LAI), irrigation, precipitation, and periodical soil water storage. It also includes daily crop evapotranspiration and seasonal crop water use assessment, harvest index, and crop yield. Hourly and daily climatic data is also provided by CoAgMET, Colorado's weather information network.

Figure 8 illustrates parameters influencing irrigation estimation in the water balance approach that we used for predictions.

6 Results and Discussion

6.1 Data Preprocessing: Principal Component Analysis (Study of Variables)

In this study, we want to use PCA as an essential step in the preprocessing of data necessary for reducing the dimension of the orthogonal variables used in linear regression and thus for reducing the number of input parameters of neural networks.

6.1.1 Normalization of Variables (Data Reduction)

In PCA the reduction of heterogeneous variables is essential when the units of measure are different. It's about working on dimensionless variables. As in our case, we have heterogeneous variables, and we proceeded to reduce them. Multiple

Table 2 Data set of the evaluated simulation results

Date	Trt_code	Growth_stage	Nitrogen_Appl (kg/ha)	...	ETc_WB (mm)	ETc_BREB (mm)	SWD_Pred_105	SWD_Pred_RZ
29/04/2012	1			...			45.0	6.0
30/04/2012	1	Plant	42	...	1.17		46.2	7.2
01/05/2012	1			...	0.61		46.8	7.8
02/05/2012	1			...	0.64		47.4	8.4
03/05/2012	1			...	0.61		48.0	9.0
04/05/2012	1			...	0.39		48.4	9.4
05/05/2012	1			...	0.19		48.6	9.6
06/05/2012	1			...	1.63		48.2	9.2
07/05/2012	1			...	1.65		47.3	8.3
08/05/2012	1			...	1.75		47.7	8.7
09/05/2012	1			...	0.57		48.2	9.2
10/05/2012	1			...	0.69		48.9	9.9
11/05/2012	1			...	1.73		48.6	9.6
12/05/2012	1			...	0.27		48.9	9.9
...	...	...		...	...	...	...	...
21/10/2012	1	Harvest		...	0.47		69.5	69.5
...	...	...	...	...	...	...	...	...
11/5/2013	12	Harvest			0.34		39.8	39.8

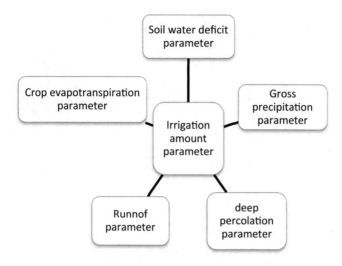

Fig. 8 Parameters influenced water balance approach

are the techniques used for variables normalization, and the highest answered one is
to divide the values by their standard deviation were calculating Euclidean distance
between individuals (12). It means that we are working on centered and reduced
variables. This new distance makes it possible to assign a more equitable role
(importance) to each of the variables> [25].

$$d^2(U_i, U_{i'}) = \sqrt{\sum_{j=1}^{p} \frac{(x_{ij} - \overline{x_{i'j}})^2}{\sigma_j^2}} \tag{12}$$

After the normalization of the data, we performed a principal component analysis of 3 components and we obtained the correlation circle, the projection of the
individuals, and the projection of the contributions of the variables to the principal
components presented in Fig. 9. It appears clearly according to the analysis of these
graphs that the most relevant variables are those projected with great dependence on
the third component present in Table 3.

The deepest colors represent the most relevant parameters guaranteeing a better
distribution of the eigenvalues. Consequently, based on the variables are reduced
from 28 to 12 while keeping the parameters necessary for irrigation water needs
estimation.

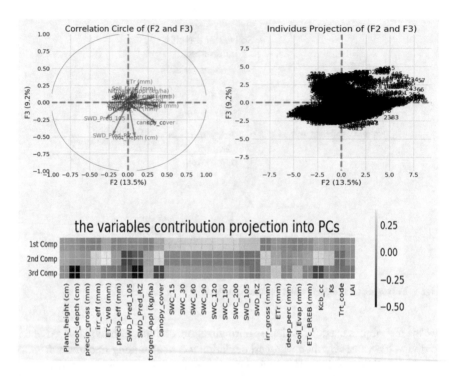

Fig. 9 Parameters influenced water balance prediction selected using PCA

6.2 Implementation

In this section, we show the obtained results after the training and testing stages of each prediction method, and we provide different measures to compare their efficiency. After selecting the relevant features, we performed predictions using XGBoost, Random Forest, and Deep Artificial Neural Networks with the configuration above, based on the same training features, training period (4074 days), and test period (The last growing season from May to November 2013:175 days). Then we estimated irrigation amount in two depths (105 cm and root zone) for two evapotranspiration estimation methods, in particular, ETc determined by experts based on Water balance (WB), FAO Penman-Monteith (FAO PM), and Bowen Ratio-Energy Balance (BREB).

For the ANN configuration, we used a sequential model with the rmsprop optimizer, an input layer of the 12 relevant features with the activation function relu, a hidden layer with the activation function relu, and an output layer, MSE as loss function, MAE, and accuracy as metrics. Figure 10 shows the ANN model used to perform features prediction.

The obtained results in Figs. 11 and 12 portray the global comparison of estimated irrigation amount between two depths using both WB, BREB, and FAO-PM approaches. It seems that the amount of irrigation estimated in the root zone is

Table 3 Daily features measurements of irrigation of maize used in the training process

Feature	Signification	Measure	Prediction state
Rooting depth	Estimated rooting depth used to estimate available soil water to the plant	(cm)	No
Plant_height	Plant height	(cm)	No
Canopy cover	The fraction of ground surface covered by the plant used to adjust Kc	–	No
Precipitation	Precipitation	(mm)	Yes
Irrigation	Irrigation amount	(mm)	Estimated
ETr	Reference evapotranspiration for a tall crop	(mm/d)	Yes
Ks	Stress coefficient; used to adjust crop coefficient based on soil water deficit	–	Yes
Kcb	Basal crop coefficient based on literature values adjusted by canopy cover	–	Yes
ETc_BREB	Crop evapotranspiration estimated by Crop evapotranspiration estimated using Bowen Ratio-Energy Balance	(mm)	Yes
ETc_WB	Crop evapotranspiration estimated using water balance	(mm)	Yes
SWD_Pred_105	Predicted soil water deficit to 105 cm depth by the water balance calculations	(mm)	Yes
SWD_Pred_RZ	Predicted soil water deficit in the current root zone by the water balance calculations	(mm)	Yes

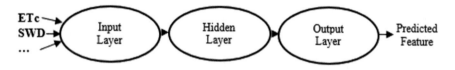

Fig. 10 The deep ANN model's configuration

closer to that estimated by the experts and the FAO-PM outperforms both WB and BREB approaches regarding precision.

Figures 13, 14, and 15 portray the comparison of estimated irrigation amount using the predictive process with the expert's estimations using both WB, BREB, and FAO-PM approaches. It seems that the amount of irrigation estimated in the root zone using FAO-PM is closer to that estimated by the experts and ANN outperforms both XGBoost and Random Forests in terms of accuracy.

From the performance analysis shown in Table 4, it appears that the ANN model outperforms both Random Forests and XGBoost models in terms of precision. Also, we can notice that estimating irrigation based on soil water deficit in the root zone using both water balance and FAO-PM approaches has improved the accuracy.

Monthly irrigation of corn (mm) in 105 cm: WB compared to BREB
and to FAO

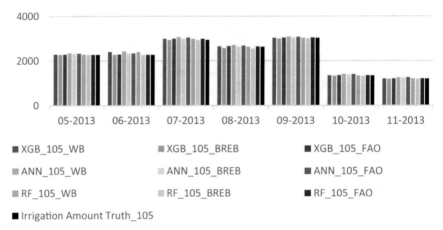

- XGB_105_WB
- ANN_105_WB
- RF_105_WB
- Irrigation Amount Truth_105

- XGB_105_BREB
- ANN_105_BREB
- RF_105_BREB

- XGB_105_FAO
- ANN_105_FAO
- RF_105_FAO

Fig. 11 Monthly estimated water requirement compared to the irrigation applied by the experts in 105 cm depth for the last growing season of corn in 2013

Monthly irrigation of corn (mm) in the Root Zone: WB compared
to BREB and to FAO

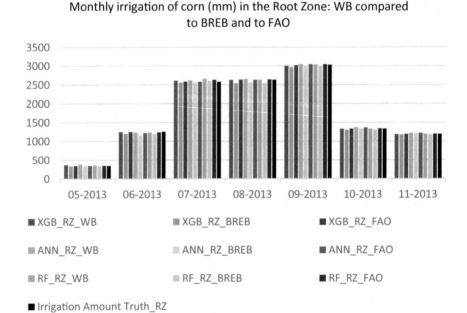

- XGB_RZ_WB
- ANN_RZ_WB
- RF_RZ_WB
- Irrigation Amount Truth_RZ

- XGB_RZ_BREB
- ANN_RZ_BREB
- RF_RZ_BREB

- XGB_RZ_FAO
- ANN_RZ_FAO
- RF_RZ_FAO

Fig. 12 Monthly estimated water requirement compared to the irrigation applied by the experts in the root zone for the last growing season of corn in 2013

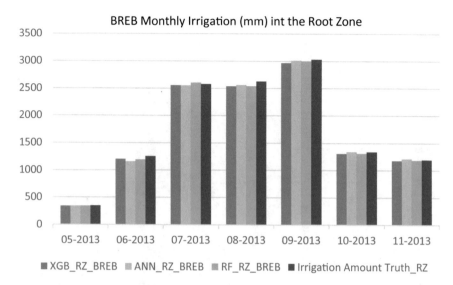

Fig. 13 Monthly estimated water requirement compared to the irrigation applied by the experts in the root zone for the last growing season of corn in 2013

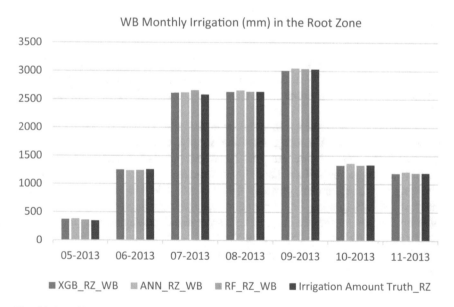

Fig. 14 Monthly estimated water requirement compared to the irrigation applied by the experts in the root zone for the last growing season of corn in 2013

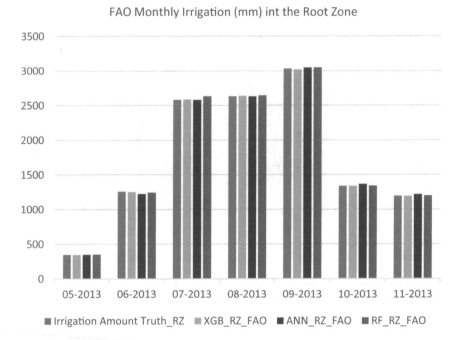

Fig. 15 Monthly estimated water requirement compared to the irrigation applied by the experts in the root zone for the last growing season of corn in 2013

In Figs. 16 and 17, we can see that the daily water requirement prediction estimated using FAO-PM follows quite precisely the tendency of the irrigation estimated by experts over time along the growing season of corn in both Root Zone and 105 cm depth.

Ultimately, meeting corn's water needs is above all knowing the water status of its soil and helping to recharge its water supply. Current methods measure the moisture in the soil at different depths. Thus, new information technologies make it possible to predict the soil state and the environment to align with climate change. Investing in this equipment, provided that it is well laid and that we know how to interpret the curves, allows us to plan and manage irrigation as well as possible. In addition, improving the accuracy of soil water status prediction and applying the appropriate water need estimation method will enhance not only the water use optimization in short and long terms but also in ensuring a save crop cultivation and consequently increasing the yield.

Table 4 Confusion Matrix of the performance analysis

Model	Depth	WB				BREB			FAO-PM		
		RMSE	MAE	R-square accuracy		RMSE	MAE	R-square accuracy	RMSE	MAE	R-square accuracy
ANN	105	11.50	5.80	0.669		11.50	4.99	0.668	11.29	5.39	0.680
ANN	**RZ**	9.61	3.85	0.838		10.44	5.29	0.809	**9.58**	**4.01**	**0.839**
XGB	105	11.47	4.33	0.670		12.00	5.88	0.639	11.29	4.30	0.681
XGB	RZ	9.77	3.92	0.833		10.57	5.52	0.804	9.61	3.90	0.838
RF	105	11.35	4.17	0.677		11.92	5.75	0.644	11.19	4.14	0.68
RF	RZ	9.75	3.84	0.834		10.56	5.43	0.805	9.61	3.80	0.83

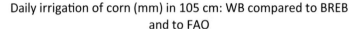

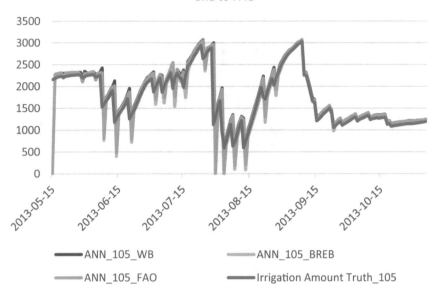

Fig. 16 Daily estimated water requirement compared to the irrigation applied by the experts in 105 cm depth for the last growing season of corn in 2013

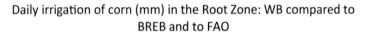

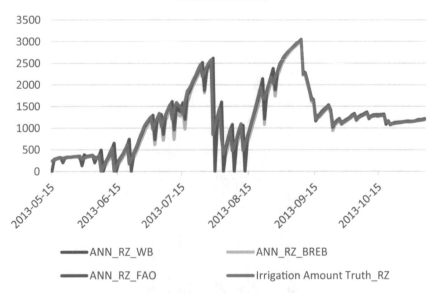

Fig. 17 Daily estimated water requirement compared to the irrigation applied by the experts in the root zone for the last growing season of corn in 2013

7 Conclusion and Perspectives

Currently, most farmers recklessly exploit water resources. These behaviors lead to the waste of a large amount of water, which will lead to resource depletion soon. Tools for calculating and predicting water needs can only be beneficial for irrigation planning and optimize the use of water resources.

This work describes a process of developing an infrastructure for forecasting crop water requirements based on three methods of estimating evapotranspiration and several methods of machine learning. This combination of methods would make it possible to have a more appropriate tool for a given context. This is a very important step towards an intelligent irrigation system.

One perspective of this work will be the implementation of a mobile system that can assist the whole farming season, from seeding to harvesting the plant. A mobile, active and strong strategy to the proposed automation outlines the future expanse. The automation of irrigation control processes using environmental parameters detected through the Internet of Things and data mining tools would increase productivity, reduce water consumption and soil erosion.

References

1. UNESCO: The United Nations World Water Development Report 2019: Leaving no one behind (2019)
2. Cisty, M., Soldanova, V.: Flow prediction versus flow simulation using machine learning algorithms. In: Lecture Notes in Computer Science (including subseries Lecture Notes in Artificial Intelligence and Lecture Notes in Bioinformatics) (2018)
3. Gupta, A., Gusain, K., Popli, B.: Verifying the value and veracity of extreme gradient boosted decision trees on a variety of datasets. In: 11th International Conference on Industrial and Information Systems, ICIIS 2016—Conference Proceedings (2016)
4. Janani, M., Jebakumar, R.: A Study on Smart Irrigation Using Machine Learning. (2019). doi: 10:23880
5. Smith, M., Allen, R., Pereira, L.: Revised FAO methodology for crop-water requirements. Int. At. Energy Agency (1998)
6. Prathibha, S.R., Hongal, A., Jyothi, M.P.: IOT based monitoring system in smart agriculture. In: Proceedings—2017 International Conference on Recent Advances in Electronics and Communication Technology, ICRAECT 2017 (2017)
7. Sentelhas, P.C., Gillespie, T.J., Santos, E.A.: Evaluation of FAO Penman-Monteith and alternative methods for estimating reference evapotranspiration with missing data in Southern Ontario, Canada. Agric. Water Manag. **97** (2010). https://doi.org/10.1016/j.agwat.2009.12.001
8. Landset, S., Khoshgoftaar, T.M., Richter, A.N., Hasanin, T.: A survey of open source tools for machine learning with big data in the Hadoop ecosystem. J. Big Data **2** (2015). https://doi.org/10.1186/s40537-015-0032-1
9. Gillick, D., Faria, A., DeNero, J.: MapReduce: Distributed Computing for Machine Learning. Icsiberkeleyedu (2006)
10. Maha, M.M., Bhuiyan, S., Masuduzzaman, M.: Smart board for precision farming using wireless sensor network. In: 1st International Conference on Robotics, Electrical and Signal Processing Techniques, ICREST 2019 (2019)

11. Ramya, R., Sandhya, C., Shwetha, R.: Smart farming systems using sensors. In: Proceedings —2017 IEEE Technological Innovations in ICT for Agriculture and Rural Development, TIAR 2017 (2018)
12. Gutierrez, J., Villa-Medina, J.F., Nieto-Garibay, A., Porta-Gandara, M.A.: Automated irrigation system using a wireless sensor network and GPRS module. IEEE Trans. Instrum. Meas. **63** (2014). https://doi.org/10.1109/TIM.2013.2276487
13. Andales, A.A., Chávez, J.L., Bauder, T.A.: Irrigation scheduling: the water balance approach. Color State Univ. Ext. (2011)
14. Buttar, N.A., Yongguang, H., Shabbir, A., et al.: Estimation of evapotranspiration using Bowen ratio method. IFAC-PapersOnLine **51** (2018). https://doi.org/10.1016/j.ifacol.2018.08.096
15. Allen, R.G., Pereira, L.S., Raes, D., Smith, M.: Crop evapotranspiration—guidelines for computing crop water requirements—FAO Irrigation and drainage paper 56 (1998)
16. Allen, R.G., Pereira, L.S., Raes, D., Smith, M.: Crop evapotranspiration: guidelines for computing crop requirements. Irrig. Drain Pap No 56, FAO (1998). https://doi.org/10.1016/j.eja.2010.12.001
17. Breiman, L.: Random forests. Mach. Learn. **45** (2001). https://doi.org/10.1023/A:1010933404324
18. Schmidhuber, J.: Deep learning in neural networks: an overview. Neural Netw. **61** (2015)
19. Gupta, N.: Artificial neural network. Netw. Complex Syst. **3**, 24–28 (2013)
20. Guerrien, M.: L'intérêt de l'analyse en composantes principales (ACP) pour la recherche en sciences sociales. Cah des Amériques Lat (2003). https://doi.org/10.4000/cal.7364
21. Ringnér, M.: What is principal component analysis? Nat. Biotechnol. **26**, 303–304 (2008)
22. Jolliffe, I.T.: Principal Component Analysis. Springer-Verlag, New York 19862 (2002)
23. Neeser, C., Dille, J.A., Krishnan, G., et al. WeedSOFT®: a weed management decision support system. Weed Sci. **52** (2004). https://doi.org/10.1614/p2002-154
24. Comas, L.: USDA-ARS Colorado Maize Water Productivity Dataset 2012–2013
25. Duby, C., Robin, S.: Analyse en composantes principales. Inst. Natl. Agron. Paris-Grignon. **80** (2006)

Machine Learning Applications for Smart Sensor Networks

Graph Powered Machine Learning in Smart Sensor Networks

Namita Shrivastava, Amit Bhagat, and Rajit Nair

Abstract A generic representation of sensor network data can be done by inherent graph structure within IoT sensor networks. We can develop a standardized graph-based framework and graphical features to support different prediction tasks from additional IoT network sensor data. In that case, we can offer IoT apps a tool for boosting prediction tasks. Graphic approaches in several IoT applications have been successful. This chapter presents the basic concepts and reasons behind a graphics learning project. It shows some basic principles to understand how graph models can be useful tools to produce better outcomes at the end of machine learning. The chapter covers the framework based on Graphical Features with some deep learning approaches such as GCN (Graph Convolutional Network) Approach, Deep Graph Convolutional Neural Network (DGCNN) and Window-Based Approach with Graphical Features. The work will also show that graphical-based features based on a smart sensor network can perform better than non-graph based features.

Keywords Sensor network · Feature selection · Neural network · Graphical features · Convolutional network

1 Introduction

Graphs are used to differentiate between objects of interest, model any simple and complex networks, or, generally speaking, depict real-life problems [1]. Since they are based on rigorous and straightforward formalism, they are applied from computer science to historical science in many science fields. We shouldn't be shocked to see them as a vital tool for intuitive properties and many useful features in Machine Learning. Graphical machine learning is rapidly a resilient logic that

N. Shrivastava · A. Bhagat
Maulana Azad National Institute of Technology, Bhopal, India

R. Nair (✉)
Jagran Lakecity University, Bhopal, India

© The Author(s), under exclusive license to Springer Nature Switzerland AG 2022   209
U. Singh et al. (eds.), *Smart Sensor Networks*, Studies in Big Data 92,
https://doi.org/10.1007/978-3-030-77214-7_9

transcends many "traditional techniques" [2]. This method is currently used by several businesses of all sizes to provide their customers with advanced learning machines.

A popular example is Google, which uses graphics machine training as a crucial part of its Expander platform: the technology behind many of the Google products and functions, for instance, in the Inbox and Smart Messaging recalls in Allo and Google Photos 15's new image recognition system. The development of a graphical machine learning framework has many advantages because graphics can be a powerful tool to solve previously mentioned problems and create advanced functionality that can be introduced without graphic support. We divide the method into four main tasks to better understand the graphs in the workflow of the Machine Learning Project:

- Data sources management involves all activities related to collecting, fusing, cleaning, and preparing the training data set for learning.
- Learning includes the use of machine learning algorithms for data set preparation.
- Store and access a template that includes predictive model storage approaches and predictive access patterns.
- Visualization that applies to all ways of data visualization to assist research.

The following mental map sums up these points:

The Fig. 1 depicted above shows the computer research project from its method point of view and is the best way to find out where we live. On the other hand, it is often useful to think about the project from a wider perspective. The following image illustrates the key points of interaction between machine learning and graphs with regard to the intent of the tasks.

The map represents in Fig. 2 conceptually shows the graphs' position in the machine learning panorama immediately. Figure 2 illustrates the map's characteristics are grouped into three main areas:

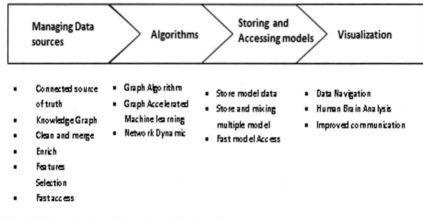

Fig. 1 Life cycle of graph based machine learning process

Fig. 2 Graph powered machine learning

- Data management: The characteristics given in this field are defined in diagrams that assist the data management learning project.
- Data Analysis: This field includes graphs-based and predictive methods.
- Data Visualization: The techniques for using graphs as a visual tool to communicate, connect with data, and find insights using the human brain are described here.

2   Literature Survey

The following section deals with the research that has already been performed in three relevant domains: Activity recognition by smart home sensors, the demographics of mobile-phone sensors, and smartphone sensors' activity recognition. Researchers have addressed the problem of movement sensor activities by using various approaches, including Bayes Belief Networks, Artificial Neural Networks [3], LogitBoost [4], Naive Bayes [5], Hidden Markov Models [6], and CRT [7]. In this regard, they have tackled the problem of moving sensor behavior detection. In addition to researching the graphical method's features, researchers have also extensively researched non-graphical strategies for improving movement detection, such as the activation of a motion sensor. Furthermore, temporal aspects of an action, such as day of the week, whether the day is a weekend, time of day, activity time, previous/next activity, and the number of motion sensor types used, are integrated into the recognition process.

Aicha et al. [8] introduced Markov Modulated Multidimensional Poison Process M3P2, which excelled over MMPP in processing two disparate heterogeneous information sources (MMPP). It was discovered that there was no correlation between the number of sensor transitions and the presence of multiple individuals or visits. Two sensors are topologically connected if activating one of them initiates an action that activates the other. One sensor may be a door sensor, so the number of sensor transitions was also considered. Brdar et al. [9] interpreted the visual representations as a graph and found a neighbor using functionality selection to evaluate work types. Mo et al. [10] went through the process of mobile phone sensor data collection, characteristics engineering, and creation of specific features for specific purposes. Using contextual knowledge, they were able to generate efficiency gains. Their characteristics are similar in their use of sensor transitions to our graphical features. The difference is that regardless of topological connections or sensor forms, they consider sensor transitions in the data. The model applies to various types of sensors and prediction jobs for multiple applications.

Mo et al. [11] illustrate the individual's mobility privacy issues because of evidence of the high degree of human trace mobility and the potential to re-identify mobility details when only a few places are used outside. The framework uses more general location-specific information, prevents complicated functionality, and uses only GPS sensor-derived graphical features.

Bouchard et al. [12] utilized longitude and raw latitude as the characteristics and addressed distance, location, shape, and gesture potential as features. The Thyme approach developed by Aminikhanghahi et al. [13] is designed to adapt a timeline based on the user activity context. Liao et al. [14] make effort to recognize GPS trace activity through training in a condition-built random field to distinguish important locations and activity marks, iteratively. They included graphic features to the characteristics used for this sector and assess the potential for enhanced performance and provide a general IoT application structure. The role of identifying behavior has recently been addressed in integrating symbolic representations of domain information and data-driven learning. Chen et al. [15] suggested a cold start, reusability, and incompleteness approach to data-driven activities modeling. In three steps, they used a hybrid ontology-based approach. In the first step, the problem of cold starts is solved through the development of seed activity models based on the information on the domain that most daily life (DL) activities are everyday routines and are typically performed in a certain area, time, and place. In the second step, activity models are generated in 2 conceptual stages, beginning with generic coarse-grained activity models accessible to all users. The reusability issue is overcome by the iterative development of individual activity models. The 3rd stage solves the incompleteness problem by updating the model behavior by using learned activity patterns.

Ye et al. [16] studied a hybrid methodology that incorporated a general ontology model with different contexts and users to reflect field knowledge. Learning methods and ontological semantics are used to unmask trends for each person's day-to-day activities. This concept has been applied to three different fields. Four approaches to activity identification in intelligent homes have been studied in visual function based on our method: GraphSVM, SubgraphSVM, and Nearest Neighbor, as well as a set of the three. Their method has been more than graphical. Hao et al. [17] proposed a real-time engine to graph concepts that can be used to classify the sequential, interlaced, and competitor human behavior patterns, all of which are correlated with enhanced human behavior detection using organized concepts analysis. Deep learning has recently been applied to the task of sensor data recognition [18]. Deep learning, like our approach, eliminates the need for manual feature design. In its classification, however, deep learning approaches can be too confident, even for inappropriate classifications, and user background knowledge is not easily removed from the network that has been studied.

In response to these problems, Rueda et al. [19] proposed the Hybrid Causal Computational Behavior Model (HCCBM). This hybrid activity recognition architecture integrates in-depth learning with symbolic models, namely CSSMs. But our graph-based representation of sensor data has not been subject to deep learning.

3 Framework Based on Graphical Features

GFF (Graphical Features-Based Framework) displays one type of sensor information in a graph, extracts graphic functions, applies techniques for selecting features, and then introduces a classification method to a prediction assignment. Figure 3 demonstrates an IoT sensor network's workflow based on the GFF for a specific prediction mission.

In Sects. 3.1–3.3, we explain the specifics of each GFF step. This article's essential contribution is to suggest and test more graphical functionality within the GFF. We speak briefly about a deep-seated approach to learning and the GCN in Sects. 4 and 4.1 to match our graphical system. GFF (Graphic Features-Based Framework) representing one type of sensor data, extracts graphic features, applies feature selection techniques [20], and then uses the prediction classification methods [21].

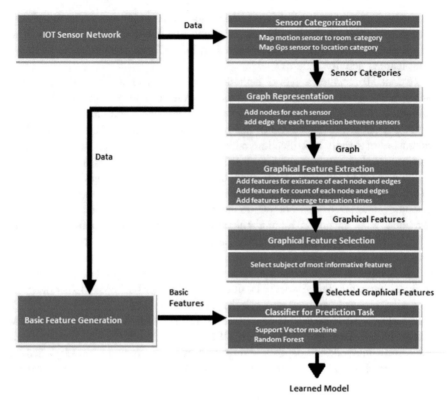

Fig. 3 IoT sensor network's workflow based on the GFF (graphical feature-based framework)

3.1 Categorize and Graph Representation of Sensor

Every sensor in the IoT framework, when given and available, will map the sensor to a sensor category and will display it as a graphic node in which IoT network sensor data is displayed. For each instance of the IoT application, we create a graph. When a labeled assignment triggers two consecutive sensors, the undirected edge between the two nodes in the graph is inserted. You can activate two sensors twice, i.e., the edge can be triggered several times. We store the count as edge attributes with many edges. The same sensor can be triggered in some applications in a row so that self-controlled data can be collected in your graphic view. Motion sensors are known to be the nodes in the intelligent home to identify activities. Everyone used as the category of the sensor [22] e.g., a bedroom, kitchen, dining room and living room.

When analyzing location data from smartphones, we believe the phones' GPS devices can reliably identify the places people visit, so location data from those devices can be used for demographic forecasts and behavior recognition [23]. Each type of position is visualized as a graphic node on our map [24]. When moving from one location to another, we may add an undirected rim connecting the two nodes. In the demographic prediction task, we create a diagram for each consumer and a diagram for every operation in the operation reconnaissance task.

3.2 Extraction of Features

We create a single graph in the sensor network data for each labeled case. We identify node and edge functions for each consumer for the demographic forecast. For the activity recognition problem, we construct one graph instance for each process. The set of characteristics consists of all nodes and edges in all graphs. Initially, the presence of nodes, the presence of boundaries, and knots and edges were three distinct graphic features. Our collection consists of all single sensor categories represented as graph nodes in all instances for the life of the nodes experiment. If a node is present in the user diagram, in the corresponding case, we set the 'ON' value to that function; otherwise, we set the 'OFF' value. We measure all the unique edges in their sensor-trajectory graph for the life of edge experiments. As the feature set for an edge experiment, we use this list of all edges. We look at the graph for each instance and track the edges' presence to create an instance. If the edge remains, the value is marked as 'ON'; if not 'OFF'. In the third experiment, we integrated the presence of knots and boundaries, which we used as an integrated function in the two previous experiments. We also measured how much each node was triggered and used as features during an event. We store how many times each edge is enabled as edge attributes and use them as graphic characteristics. For the demographic prediction problem, we tried to add more transitional features. A sensor type is defined as an attribute of a node, and time for edge transfer is the time needed to move from one sensor to another. To judge their effect on task success prediction, we use the node and border attributes as more graphical characteristics and assign them these descriptive terms.

3.3 Feature Selection and Classification

Compared with the number of cases, the number of features can be high as all graphs are used to set feature nodes and edges. Feature selection is used to reduce the number of features in this state to prevent overfitting [25].

Two main approaches to pick useful and pertinent attributes can be found in the literature: filters and wrappers [26]. According to each function, filters are based on various steps, such as knowledge gain, correlation, benefit ratio, and symmetric uncertainty. The wrapper approach tests subsets with a classification algorithm and provide an optimal and chosen subset of features for the subset that performs most. The wrapper method could be computationally costly if the set's size is massive, with thousands of features. A hybrid approach is used for a large number of features when featuring sets in a low-size system are built based on the filtering method, and the wrapper approach is then used to find an optimal subset of features. On this optimally selected set of features, we then apply a suitable classification technique. We also found the Vector Support Machine and various tree-based classifications for the classifiers. We explain each application's specifics in the section Results of the feature selection algorithms and the classification methods that produced the best results.

3.4 Additional Features

There are additional features such as uniform counts of nodes, edge counts and, edge transfer times. These additional features are presented in the smart home domain. We also show categorical information for intelligent home sensors using the sensor type and room type. We conducted an intelligent home experiment to determine the effect of various window sizes on graphics in activities classification.

4 Deep Learning

Deep Learning (DL) enables us to construct relevant features without the user interaction. We, therefore, use it over the Graphical Feature-based Framework (GFF) [27]. A DL network can detect related features and prevents users from manually developing them. If the DL classification can build these features internally, then adding graphical features can also be eliminated. To evaluate this, we include a DL classifier with raw sensor data and graphical features. We then equate the outcomes with our GFF.

4.1 GCN (Graph Convolutional Network) Approach

We also research the application in an intelligent home by a Graph Convolutional Network (GCN). The GCNs have developed several methods to redefine the graph data convolution with the first popular research presented in Bruna et al. [28]. In computer vision, recommendation systems, traffic network forecasting and molecular chemistry [29], graph-based neural networks (GNNs) have been implemented. Zhang et al. [30] proposed approach called the Deep Graph Convolutionary Neural Network (DGCNN), which takes a chart directly as input for a classification function and thus eliminates the need to turn the chart into a vector. On many benchmark data sets, the approach was better than the current methods. DGCNN is applied to the sensor data graphs in the intelligent home domain.

4.1.1 Advantages of Graph Convolutional Network

To better understand the advantages of GCN, we have shown the comparison of the GCN with the two existing approaches, a support vector machine (SVM) [31] that employs various input variables to evaluate building groups and a random forest (RF), which mainly focuses on Gestalt concepts of proximity, continuity, and integration [32].

Input variables for the GCNN method were the area, SBRO, compactness, and concavity indices. In addition to the three building-group-specific features already mentioned (total edge weight, the smallest bounding rectangle, and the building's ratio to this smaller bounding rectangle), three additional features are extracted: edge weight standard deviation, the area difference, and the orientation difference for the SVM and Random forest. One part of the Guangzhou dataset was used as a training dataset, and the other part of the Guangzhou with complete Shanghai dataset was used as a testing dataset [33]. Below Table 1 summarizes the findings in terms of training and testing accuracy with the comparison of the GCNN model and the other algorithms'.

The Guangzhou dataset shows that all three classification methods achieve acceptable classification accuracy. The GCNN was accurate to 98.02% when only four indexes were used, which is better than the two other approaches. A shift in the data's size or geographic reach does not impact the features that can be extracted. It is possible to conclude that the GCNN outperforms the other ways by analyzing the classification and generalization potential.

Machine learning approaches for classifying building group patterns have not automated the classification process because they manually define group units' functions. Because of these handcrafted features, the final classification results have a considerable effect, and a wealth of imagination and extensive experience is required to design them. As the amount of data increases, this process becomes more complicated and unpredictable.

A highly customized GCNN that is trained and learned based on an individual's input variables and connection modes is proposed. A function extractor (convolution kernel) is used in the filter, and parameters for these kernels are learned from examples with labels. An end-to-end approach covers the entire process. A further benefit of the suggested approach is that it requires a single processing unit, which is better in line with human cognition. Some of the applications of GCNN are as follows:

Table 1 Comparison of GCNN with SVM and random forest

Method	Shanghai dataset test accuracy (%)	Guangzhou dataset test accuracy (%)
GCNN	75.59	92.93
Random forest	87.75	95.14
Support vector machine	92.60	98.02

(i) Natural Language Processing—One of the oldest and most traditional NLP strategies is the classification of text. The citation network in which attributes of documents are primarily modeled using the bag-of-words may be built. Classification of documents as superficial nodes is the correct way in this situation. To perform the classification of text, other than multiple graphs, convolutional network models have been proposed. In addition, instead of displaying documents individually and classifying the texts based on classification at the document level, you can visualize all the documents as a graph and classify the texts by graph classification. Although it is not explicitly mentioned, the TextGCN [34] models a whole corpus, applies word embedding and document embedding together, and then trains a softmax classifier to classify text. Gao et al. [35] combine the node ordering information from a graph pooling layer with hybrid convolutions, which results in a more efficient network than the conventional CNN- and GCN-based methods. These single-granularity methods can provide a suboptimal output when there are several different labels at varying detail levels. The convolutional graph network plays a central role in developing various NLP-related applications, as it was found to be the case with knowledge extraction.

A good example is where recurrent neural networks' machine learning technique is used to produce context-aware hidden representations of words or sentences. Then an encoder and decoder are employed to mark the words at the word level. Although knowledge extraction, such as named entity extraction, can be achieved using GraphIE, other extraction forms cannot. This research (along with several others) indicates that Zhang et al. [36] findings may be accurate. A convolutional graph network has been developed to compare word-event relations and word-event correlations [37]. Marcheggiani et al. [38] build on the findings of prior syntactic dependency tree-based NLP models and the methodology built-in that research may be useful for applications like semantic function labeling and neural machine translation [39]. A semantic bias is injected into sentence encoders using graph convolutional networks, and the output gains can be observed [40]. It is planned for dependency parsing.

(ii) Computer Vision—Since the early twentieth century, the use of computers in computer vision and robotics has been growing. Convolutional neural networks, or CNNs, have been adopted for image recognition tasks, especially in object recognition (CNNs). Despite CNNs' considerable accomplishments, it isn't easy to include the project's intrinsic graph structures into the application domain. Convolutional neural networks have been used to solve computer vision problems, and their efficiency was found to be similar or even better. This is in addition to the other classification algorithms, which distinguishes these applications by the form of data they manipulate [41].

(iii) Science—When particle physicists speak about jets, they are referring to the intensely concentrated sprays of energetic hadrons. In other words, when researchers discuss jets, they are talking about collimated sprays of energetic hadrons. This type of neural network was recently developed to classify jets into two groups: those based on quantum chromodynamics (QCD) and W bosons (W bosons). ParticleNet, based on edge convolutions, processes jet clouds directly to tag particles [42]. Although all this has been achieved for IceCube data classification, convolutional graph-based network models have also been applied to the IceCube signal classification. A yet-undeveloped application uses the results to quantify the physical dynamics, such as the deformation of a cube as it reaches the earth. To investigate an object from the ground up, it is best to break it down into individual particles and then bind them together in the hierarchical tree-like structure suggested by Mrowca et al. [43].

Drug discovery, materials science, and chemistry have seen enormous development because of learning about molecules. Furthermore, convolutional neural networks have been employed to predict molecular fingerprints [44]. An attention-based graph convolutional network model, is employed for chemical stability prediction of a compound named as DeepChemStable [45]. Polypharmacy side effects can be predicted using graph convolutions. Molecular property prediction is just as important in other fields as well. Message-Passing Neural Networks (MPNNs) can be used to simulate quantum properties of a molecular by using the above definition [46]. PotentialNet, for the first time, includes graph convolutions that learn the properties of atoms [47]. It continues by running a propagation technique that maps chemical bonds into spatial distances, and later, by collecting and processing graphs on the ligand atoms. It ends with a completely connected layer for molecular property predictions. It isn't easy to estimate the position of protein interfaces. This is an issue with sensitive applications of drug discovery. In a graph created by Fout et al. [48], each amino acid residue in a protein is represented as a node and provides various information about the sequence, structure, and characteristics. Different protein graphs are graph convolution layers in which protein binding sites and sites of interaction between proteins are graphed, followed by one or more completely connected layers to predict protein interface. Furthermore, appears to learn material properties by directly measuring interactions between atoms in the crystal.

(iv) Social Network Analysis—GCNN has also been used in other areas, including group detection [49] and link prediction [50]. DeepInf [51] plans to use awareness of the users' latent features to forecast social factors. This team wants to use graph convolutional networks to predict the amount of retweets following a new article. In addition, fake news can be identified by convolutional neural networks [52]. Graph convolutional networks have found applications in several application fields, such as social recommendation, i.e., recommendations based on interactions between users and objects. PinSage [27] exploits Pinterest interactions between pins and boards by using GraphSAGE [53]. Wang et al. used a method that uses the inter-actions between users and objects in a neural graph collaborative fltering system and takes advantage of these interactions when implementing the graph convolutional network.

4.1.2 DGCNN (Deep Graph Convolutional Neural Network)

Most of the current models still suffer from shallow structures, though the deeper architecture is being used for representation learning. It might not be advantageous to use more graph convolution layers when using GCN in practice only uses two layers. The quick method of propagating makes it all the more intuitive. The further apart nodes are, the smoother their representations become, except for distinct and far-flung nodes. It is in direct contradiction to the reason why you'd use a deep model in the first place. Although no concrete work has been proposed to solve this problem (i.e., use skip link models), the task of creating a more scalable, deep adaptive architecture for graphs is still open.

While there has been a good deal of advancement in the area, most previous studies concentrate on shallow GCNs while the more nuanced and significant extension is rarely addressed. Residual GCNs were first proposed with a residual mechanism in the back of the mind [54]. It turns out, as discovered in their experiments, residual GCNs still perform worse than expected when the depth is three and beyond. Li et al. [55] describe the most significant barrier to deeper networks: over-smoothing. Unfortunately, the article fails to provide any ideas about how to solve this issue. After successfully smoothing out over-smoothing in the initial analysis, a follow-up study Klicpera et al. [56] employs customized PageRank addition to introducing the rooted node in the message passing loop. However, accuracy still degrades when the depth increases from 2. To per-form multi-hop message passing, Xu et al. [57] implements dense connections. It also uses DropEdge, which is compatible with deep GCNs, to formulate them. According to Oono and Suzuki [58], GCN nodes will converge to a subspace and lose details. GCNs incorporating residual layers, dense connections, and dilated convolutions are now easier to create thanks to the recent progress. Although this model is only applicable to graph-level classification (i.e., point cloud segmenta-tion), the data points used in the model aren't disconnected like how they appear in a graph. In the task for node classification, all of the samples are nodes, and they're all connected, so we must mitigate the over-smoothing problem. The over-smoothing reduction can be achieved by using DropEdge, and enhanced deep GCNs for classifying nodes.

Many convolutional neural networks use static input graphs; this principle holds for current graph convolutional networks. The reality is, however, in the real world, networks are continually evolving. For example, social networks are dynamic networks because users enter and leave them regularly, and because of this, interactions between users are always changing. Although doing so may not produce the most desirable results, learning graph convolutional networks on static graphs cannot give optimal output. Even though these dynamic graph convolutional networks are an efficient dynamic graph analysis method, they must be studied further.

5 Window-Based Approach with Graphical Features

The discussion in the previous section is on the creation of graphs, nodes, and edges. Now, break down the task into smaller bits to observe precisely how the window size influences behavior detection. We have no idea during an operation instance when the operation instance begins or finishes, so sensor data is collected during a specified time frame. Using the given window size, the different operations are divided into pieces based on the window's size and used to generate bar graphs in those segments. We collect the minimum, maximum, and average window sizes for each data set method to change the window sizes range.

Window size varied from 5 s to the average window size for each dataset (Aruba 1895 s, Cairo 1020 s, Tulum 4478 s). A 10-fold cross-validation SVM is used (using Scikit Learn LinearSVC). Instead of breaking down tasks into windows of a certain duration, "Complete activity" is used as a measure of graphical characteristics [59]. For each node, edges, and nodes, the complete operation is represented using a horizontal line.

6 Conclusion

In order to enhance prediction tasks efficiency we researched and tested graphical depictions and graphical functions on the IoT sensor networks. In order to apply sensor network data, graphical feature based framework (GFF) is discussed. This kind of system is structured and used in a multiple way. First of all, the system uses a Graph structure inherent to the sensor network data. Secondly, the Architecture provides a broad approach to using graphical features to boost prediction accuracy across various sensor networks. Thirdly, the system strengthens predictions without implementing complex application and prediction tasks.

In this work, GCNN is discussed along with its application in various domains. The chapter has also shown how a GCNN can performs better than existing machine learning algorithms such as SVM and Random forest in terms of classification accuracy. In addition chapter has also covers the Deep Graph Convolutional Neural Network (DGCNN). Overall this work has shown the importance of Graph Convolutional Network in the analysis world.

7 Future Research

This review offers some recommendations for future studies. One research direction consists of extracting multi-edge paths and little subscribers along with knots and edges as functionality and evaluating their effects on IoT network device prediction tasks. The two-way transitions from motion sensors in the smart home on a single testbed yielded extracted features. Increasing the number of paths and subgraphs while also increasing the number of functions can be time- and resource-intensive for computers. But, before using multi-edge feature selection as a route, it's essential to carry out a consistent exploration of multi-edge transitions, which will help shed light on whether multi-edge feature selection features are a promising direction. We also want to compare our emphasis and assess the possibilities of integrating methods for enhancing sensor network prediction tasks with symposium-based representation approaches discussed in the related work section. The framework of high-level activities is used for symbolic systems and hidden Markov models, and our method takes advantage of the sensor technology. We can analyze whether the structures found in both cases are similar and whether the combination of high-activity structure and low sensor structure is beneficial to the model.

We aim to boost the prediction task efficiency of classifiers other than SVM and Random Forest and assess GFF. We want to discuss why the various classifiers perform best with multiple datasets and applications. On this basis, we want to provide the best practice for the selection of classifiers from the GFF. We may turn them into nominal features for continuous features, such as nodes and edge counts. Dichotomization is called the transformation of a continuous variable into a binary variable. The presence of knots and edges is a dichotomization of the counts of knots and edges in our work. Dichotomized features can lead to data loss, and performance decreases.

Another problem is the choice of the cutting point for dichotomizing the function. The amount of information lost depends on cutting issues, and the optimal cutting point depends on the unknown parameters [45]. This is in line with our results in the smart home setting, in which we have better output than life with counting features. However, many of the algorithms for feature selection give distinct characteristics of better performance.

Discretization is used as a pre-processing step in the selection of correlation-based features. This is consistent with our demographic forecast in the GPS data, in which features of over 7000 are needed to boost performance to select fewer, more essential features. In this case, it was useful to discrete features before giving them an algorithm for the function collection. Eventually, we will use methods described in this work to experiment with discretizing counts of nodes and edges in intervals to minimize information loss and to use the discretized features simultaneously. Usually, forecasters model future events based on historical data and time-series data. Cross-validation of random data sub-sets thus is a bad option because we can quickly learn to look at both the details we expect to forecast both before and after, which is different from the actual life situation, in which we need to predict future events based on past events.

We will use test data that occurs later in a dataset instead of randomly-selected subsets to explore time series forecasting techniques for IoT sensor network data. The general pattern that human activities adopt is of visited nodes and transitions on edge. The way people act changes as they go through life, and the hidden test data that was not seen in the training data will potentially include new transitions. To evaluate the effect of these new transitions on active recognition accuracy, we can evaluate the models' performance drift on the new test data. Models should be retrained if the amount of performance decreases over time.

Although comprehensive learning approaches to sensor data recognition have been promising, they have performed less well in our environment than our non-depth learning. However, substantial networking will finally mean that our architecture is built to escape the effort of developing features in conventional machine learning methods. In other IoT cases, commonly used for tracking and monitoring, the utility of the GFF can be evaluated. For three different domains, we have developed three various applications to process and model data. To provide sensor information as part of this input interface, we will create a single software application and establish an overall interface for all IoT data.

References

1. Monti, C., Boldi, P.: Estimating latent feature-feature interactions in large feature-rich graphs. Internet Math. (2017)
2. Pattern recognition and machine learning. J. Electron. Imaging (2007)
3. Kwon, S.J.: Artificial neural networks (2011)
4. Kotsiantis, S.B., Pintelas, P.E.: Logitboost of simple Bayesian classifier. In: Informatica (Ljubljana) (2005)
5. Ye, N., Ye, N.: Naïve Bayes classifier. In: Data Mining (2020)
6. Karlof, C., Wagner, D.: Hidden Markov model cryptanalysis. Lect. Notes Comput. Sci. (including Subser. Lect. Notes Artif. Intell. Lect. Notes Bioinformatics (2003)
7. Bland, C., et al.: CRISPR recognition tool (CRT): a tool for automatic detection of clustered regularly interspaced palindromic repeats. BMC Bioinf. (2007)
8. Aicha, A.N., Englebienne, G., Kröse, B.: Modeling visit behaviour in smart homes using unsupervised learning. In: UbiComp 2014—Adjunct Proceedings of the 2014 ACM International Joint Conference on Pervasive and Ubiquitous Computing (2014)
9. Brdar, S., Ćulibrk, D., Crnojević, V.: Demographic attributes prediction on the real-world mobile data. In: Proceedings of Mobile Data Challenge by Nokia Workshop, in Conjunction with International Conference on Pervasive Computing (2013)
10. Mo, K., Tan, B., Zhong, E., Yang, Q.: Report of task 3 : your phone understands you. Mob. Data Chall. Work. (2012)
11. Hu, J., Zeng, H.J., Li, H., Niu, C., Chen, Z.: Demographic prediction based on user's browsing behavior. In: 16th International World Wide Web Conference, WWW2007 (2007)
12. Bouchard, K., Holder, L., Cook, D.J.: Extracting generalizable spatial features from smart phones datasets. In: AAAI Workshop—Technical Report (2016)

13. Aminikhanghahi, S., Fallahzadeh, R., Sawyer, M., Cook, D.J., Holder, L.B.: Thyme: improving smartphone prompt timing through activity awareness. In: Proceedings—16th IEEE International Conference on Machine Learning and Applications, ICMLA 2017 (2017)

14. Liao, L., Fox, D., Kautz, H.: Location-based activity recognition using relational markov networks. In: IJCAI International Joint Conference on Artificial Intelligence (2005)

15. Chen, L., Nugent, C., Okeyo, G.: An ontology-based hybrid approach to activity modeling for smart homes. IEEE Trans. Human-Machine Syst. (2014)

16. Ye, J., Stevenson, G., Dobson, S.: USMART: an unsupervised semantic mining activity recognition technique. ACM Trans. Interact. Intell. Syst. (2014)

17. Hao, J., Bouzouane, A., Gaboury, S.: Complex behavioral pattern mining in non-intrusive sensor-based smart homes using an intelligent activity inference engine. J. Reliab. Intell. Environ. (2017)

18. Chen, K., Zhang, D., Yao, L., Guo, B., Yu, Z., Liu, Y.: Deep learning for sensor-based human activity recognition: overview, challenges and opportunities. arXiv (2020)

19. Rueda, F.M., Ludtke, S., Schroder, M., Yordanova, K., Kirste, T., Fink, G.A.: Combining symbolic reasoning and deep learning for human activity recognition. In: 2019 IEEE International Conference on Pervasive Computing and Communications Workshops, PerCom Workshops 2019 (2019)

20. Chandrashekar, G., Sahin, F.: A survey on feature selection methods. Comput. Electr. Eng. (2014)

21. Tharwat, A.: Classification assessment methods. Appl. Comput. Informatics (2018)

22. Laurijssen, D., Truijen, S., Saeys, W., Daems, W., Steckel, J.: An ultrasonic six degrees-of-freedom pose estimation sensor. IEEE Sens. J. (2017)

23. Elliot, N.B., Cushman, S.A., Macdonald, D.W., Loveridge, A.J.: The devil is in the dispersers: predictions of landscape connectivity change with demography. J. Appl. Ecol. (2014)

24. Song Suzhou, J.S.T.V.E., Xin Suzhou, Z.S.T.V.E., Ding Suzhou, W.S.T.V.E.: Research on android intelligent phones controlling the car to run. TELKOMNIKA Indones. J. Electr. Eng. (2013)

25. Nair, R., Bhagat, A.: Feature selection method to improve the accuracy of classification algorithm. Int. J. Innov. Technol. Explor. Eng. (2019)

26. Nair, R., Bhagat, A.: A life cycle on processing large dataset—LCPL. Int. J. Comput. Appl. (2018)

27. Ying, R., He, R., Chen, K., Eksombatchai, P., Hamilton, W.L., Leskovec, J.: Graph convolutional neural networks for web-scale recommender systems. In: Proceedings of the ACM SIGKDD International Conference on Knowledge Discovery and Data Mining (2018)

28. Bruna, J., Zaremba, W., Szlam, A., LeCun, Y.: Spectral networks and deep locally connected networks on graphs. In: 2nd International Conference on Learning Representations, ICLR 2014—Conference Track Proceedings (2014)

29. Wu, Z., Pan, S., Chen, F., Long, G., Zhang, C., Yu, P.S.: A comprehensive survey on graph neural networks. arXiv (2019)

30. Zhang, M., Cui, Z., Neumann, M., Chen, Y.: An end-to-end deep learning architecture for graph classification. In: 32nd AAAI Conference on Artificial Intelligence, AAAI 2018 (2018)

31. Liqiang, Z., Hao, D., Dong, C., Zhen, W.: A spatial cognition-based urban building clustering approach and its applications. Int. J. Geogr. Inf. Sci. (2013)

32. Hecht, R., Meinel, G., Buchroithner, M.: Automatic identification of building types based on topographic databases—a comparison of different data sources. Int. J. Cartogr. (2015)

33. Yan, X., Ai, T., Yang, M., Yin, H.: A graph convolutional neural network for classification of building patterns using spatial vector data. ISPRS J. Photogramm. Remote Sens. (2019)

34. Yao, L., Mao, C., Luo, Y.: Graph convolutional networks for text classification. In: 33rd AAAI Conference on Artificial Intelligence, AAAI 2019, 31st Innovative Applications of Artificial Intelligence Conference, IAAI 2019 and the 9th AAAI Symposium on Educational Advances in Artificial Intelligence, EAAI 2019 (2019)

35. Gao, H., Wang, Z., Ji, S.: Large-scale learnable graph convolutional networks. In: Proceedings of the ACM SIGKDD International Conference on Knowledge Discovery and Data Mining (2018)
36. Zhang, N., et al.: Long-tail relation extraction via knowledge graph embeddings and graph convolution networks. In: NAACL HLT 2019–2019 Conference of the North American Chapter of the Association for Computational Linguistics: Human Language Technologies—Proceedings of the Conference (2019)
37. Liu, X., Luo, Z., Huang, H.: Jointly multiple events extraction via attention-based graph information aggregation. In: Proceedings of the 2018 Conference on Empirical Methods in Natural Language Processing, EMNLP 2018 (2020)
38. Marcheggiani, D., Titov, I.: Encoding sentences with graph convolutional networks for semantic role labeling. In: EMNLP 2017—Conference on Empirical Methods in Natural Language Processing, Proceedings (2017)
39. Bastings, J., Titov, I., Aziz, W., Marcheggiani, D., Sima'an, K.: Graph convolutional encoders for syntax-aware neural machine translation. In: EMNLP 2017—Conference on Empirical Methods in Natural Language Processing, Proceedings (2017)
40. Marcheggiani, D., Bastings, J., Titov, I.: Exploiting semantics in neural machine translation with graph convolutional networks. In: NAACL HLT 2018–2018 Conference of the North American Chapter of the Association for Computational Linguistics: Human Language Technologies—Proceedings of the Conference (2018)
41. Johnson, J., Gupta, A., Fei-Fei, L.: Image generation from scene graphs. In: Proceedings of the IEEE Computer Society Conference on Computer Vision and Pattern Recognition (2018)
42. Qu, H., Gouskos, L.: Jet tagging via particle clouds. Phys. Rev. D (2020)
43. Mrowca, D., et al.: Flexible neural representation for physics prediction. In: Advances in Neural Information Processing Systems (2018)
44. Duvenaud, D., et al.: Convolutional networks on graphs for learning molecular fingerprints. In: Advances in Neural Information Processing Systems (2015)
45. Li, X., Yan, X.. Gu, Q., Zhou, H., Wu, D., Xu, J.: DeepChemStable: chemical stability prediction with an attention-based graph convolution network. J. Chem. Inf. Model. (2019)
46. Gilmer, J., Schoenholz, S.S., Riley, P.F., Vinyals, O., Dahl, G.E.: Neural message passing for quantum chemistry. In: 34th International Conference on Machine Learning, ICML 2017 (2017)
47. Feinberg, E.N., et al.: PotentialNet for molecular property prediction. ACS Cent. Sci. (2018)
48. Fout, A., Byrd, J., Shariat, B., Ben-Hur, A.: Protein interface prediction using graph convolutional networks. In: Advances in Neural Information Processing Systems (2017)
49. Levie, R., Monti, F., Bresson, X., Bronstein, M.M.: CayleyNets: graph convolutional neural networks with complex rational spectral filters. IEEE Trans. Signal Process. (2019)
50. Kipf, T.N., Welling, M.: Variational graph auto-encoders 1 a latent variable model for graph-structured data. NIPS Work. (2016)
51. Qiu, J., Tang, J., Ma, H., Dong, Y., Wang, K., Tang, J.: DeepInf: social influence prediction with deep learning. In: Proceedings of the ACM SIGKDD International Conference on Knowledge Discovery and Data Mining (2018)
52. Monti, F., Frasca, F., Eynard, D., Mannion, D., Bronstein, M.M.: Fake news detection on social media using geometric deep learning. arXiv (2019)
53. Hamilton, W.L., Ying, R., Leskovec, J.: Inductive representation learning on large graphs. In: Advances in Neural Information Processing Systems (2017)
54. Kipf, T.N., Welling, M.: Semi-supervised classification with graph convolutional networks. In: 5th International Conference on Learning Representations, ICLR 2017—Conference Track Proceedings (2017)
55. Li, Q., Han, Z., Wu, X.M.: Deeper insights into graph convolutional networks for semi-supervised learning. In: 32nd AAAI Conference on Artificial Intelligence, AAAI 2018 (2018)
56. Klicpera, J., Bojchevski, A., Gunnemann, S.: Predict then propagate: graph neural networks meet personalized pagerank. arXiv (2018)

57. Xu, K., Li, C., Tian, Y., Sonobe, T., Kawarabayashi, K.I., Jegelka, S.: Representation learning on graphs with jumping knowledge networks. In: 35th International Conference on Machine Learning, ICML 2018 (2018)
58. Oono, K., Suzuki, T.: On asymptotic behaviors of graph CNNs from dynamical systems perspective. arXiv (2019)
59. Akter, S., Holder, L.: Improving IoT predictions through the identification of graphical features. Sensors (Switzerland) (2019)

Author Index

Printed in the United States
by Baker & Taylor Publisher Services